DR. ASH PACHAURI, PhD
&
DR. SAROJ PACHAURI, MD, PhD, DPH

AI in Education: Empowering Educators, Inspiring Students

Practical Strategies to Transform Teaching and Improve Learning Outcomes

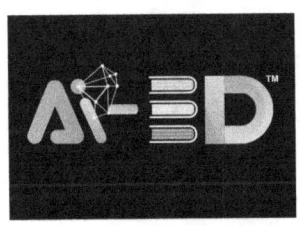

Contents

1

Introduction

Setting the Stage for a Transformative Journey in Education Through AI

Imagine the impact of stepping into a classroom where every student is deeply engaged, lessons are tailored to individual needs, and administrative tasks are effortlessly managed. Envision a learning environment where technology and human ingenuity coexist harmoniously, empowering educators to focus more on nurturing creativity, critical thinking, and a love for learning. This is not a distant dream but a burgeoning reality, thanks to the practical benefits of Artificial Intelligence (AI) in the classroom.

AI is revolutionizing industries worldwide, and education is no exception. From personalized learning experiences to automated administrative tasks, AI offers unprecedented opportunities to enhance teaching and learning. However, the journey to integrating AI into education can seem daunting. This book aims to demystify AI, making it an accessible and powerful tool for educators of all backgrounds and skill levels.

Introducing the Authors' Passion and Purpose

Our dedication to education and technology is not just professional but deeply personal. We've spent years in the classroom, witnessing the daily challenges teachers face—overwhelming workloads, diverse student needs, and the constant pressure to innovate. These experiences have not only driven us to explore how AI can alleviate these burdens and unlock new possibilities in education but have also given us a unique perspective on the practical application of AI in real educational settings.

We have seen AI's transformative potential in education through extensive research, collaboration with educators and technologists, and practical experimentation. We aim to share this knowledge, providing educators with the tools and confidence to harness AI effectively. We believe that when used thoughtfully and ethically, AI can empower educators, enhancing the human elements of education and fostering environments where students and teachers thrive.

The Book's Mission

The primary mission of this book is to demystify AI for educators and provide a practical, step-by-step guide to harnessing AI in their classrooms. We will not only explore theoretical insights and real-world examples but also provide actionable strategies and ethical considerations to effectively implement AI tools and techniques. Whether you are a seasoned tech enthusiast or a complete novice, this book is designed to meet you wherever you are in your AI journey and provide you with the practical knowledge and confidence to integrate AI into your teaching practices.

By the end of this guidebook, you will have a comprehensive understanding of AI's role in education, practical skills to integrate AI tools into your teaching, and a roadmap for continuous learning and innovation. More importantly, you will become empowered to innovate in your teaching practices, leveraging AI to enhance student engagement, personalize learning, and streamline administrative tasks.

The Book's Structure

This book is designed to provide a blend of theoretical concepts and practical tools tailored to meet the needs of educators at all skill levels. Each chapter progresses from the previous one, gradually strengthening your understanding of AI and its applications in diverse educational settings. Here is a brief overview of what to expect so you can navigate the content according to your specific needs and interests:

Chapter 1: Demystifying AI for Educators—This chapter will break down complex AI terminology into simple, understandable terms. It will provide a solid foundation, introducing key concepts and tracing the evolution of AI in education.

Chapter 2: Ethical Considerations in AI — Understanding the ethical implications of AI is crucial. This chapter will explore data privacy, bias, transparency, and accountability issues and offer guidelines for responsible AI use.

Chapter 3: AI Integration Strategies — This chapter covers practical strategies for incorporating AI into lesson plans, personalized learning pathways, and administrative tasks. We will provide step-by-step guides and real-world examples.

Chapter 4: AI for Classroom Management and Efficiency - This chapter will focus on tools and techniques to streamline classroom management, including AI chatbots, automated grading, and classroom analytic tools.

Chapter 5: Subject-Specific AI Applications - We will explore how AI can enhance learning in specific subjects, such as STEM, language, history, and art, with practical examples and success stories.

Chapter 6: Engaging Students with AI—Discover how AI can gamify learning, create immersive experiences with virtual and augmented reality, and support interactive experiments.

Chapter 7: Addressing Diverse Learning Needs with AI - This chapter will focus on AI tools that support struggling students, English language learners, and students with special needs, ensuring inclusive education for all.

Chapter 8: AI for Professional Development - Learn how AI can personalize professional learning for educators, foster peer-to-peer collaboration, and keep you up-to-date with the latest AI trends, developments, and applications.

Chapter 9: Understanding Machine Learning and Natural Language Processing - Dive deeper into advanced AI technologies, including machine learning tools and application of neural networks and natural language processing, and their educational implications.

Chapter 10: Preparing for the Future of AI in Education - Explore future trends in AI-driven personalized learning, ethical considerations, and the evolving role of educators in an AI-enhanced classroom.

Chapter 11: Building an AI-Ready Classroom - Practical advice on assessing infrastructure needs, overcoming budget constraints, fostering innovation, and ensuring safety and security in AI-enabled educational environments.

Chapter 12: Case Studies and Real-World Examples - We will conclude with inspiring case studies and real-world examples of AI integration in various educational contexts, highlighting successes, challenges, and lessons learned.

Bonus 1: Spotlight on AI in Education and Gender Equity—This bonus section explores how AI can be leveraged to promote gender equity in education by identifying and mitigating inherent biases in learning materials and assessment tools. Using AI to foster inclusive and equitable education opportunities, we can ensure that all students have fair and equal opportunities to succeed and thrive.

Bonus 2: Spotlight on AI in Education and Trust - In this bonus section, we delve into the crucial aspect of building trust in AI within the educational landscape, emphasizing the importance of transparency, ethical practices, and data security. By fostering trust, educators, and institutions can ensure that AI technologies are effectively and responsibly integrated, ultimately enhancing the learning experience for all students.

Bonus 3: Spotlight on AI in Education and Human Security – Related to and in continuation of Bonus 2, this section spotlights how AI in education impacts human security, emphasizing the need to address privacy, algorithmic bias, mental health, economic stability, and cybersecurity. By implementing comprehensive strategies to mitigate these risks, we can ensure that AI enhances education while safeguarding individual and collective well-being.

Message to the Reader

Dear Educator,

Whether you are an experienced tech user or new to AI, this book is for you. We understand that the journey to integrating AI into your teaching practices can be filled with uncertainty and challenges. We aim to provide clear, jargon-free explanations and actionable strategies to make this journey as smooth and rewarding as possible.

We recognize the diverse levels of familiarity with AI among educators. Some may already use AI tools in their classrooms, while others may be skeptical or unsure where to start. This book is designed to meet you wherever you are, offering step-by-step guidance, practical examples, and real-world success stories.

Unique Value of this Book

What sets this book apart from other resources on AI in education is its comprehensive coverage and practical focus. We delve into subject-specific applications, ethical considerations, and future trends, providing a holistic view of AI's educational potential. But we don't stop there—our primary focus is on how you, as an educator, can practically and ethically use AI to improve teaching and learning.

We emphasize actionable strategies and real-world examples, ensuring that you can apply the concepts and tools discussed in your classroom. Each chapter is designed to build your confidence and competence, transforming AI from a daunting concept into a powerful ally in your teaching toolkit.

Specific Benefits to the Reader

As you read through this book, you will gain several specific benefits:

1. Enhanced Understanding of AI: You will gain a strong understanding of AI concepts, terms, and technologies, demystifying the complex language of AI.

2. Practical Skills: Learn how to integrate AI tools into your teaching practices, from lesson planning to personalized learning and classroom management.

3. Ethical Awareness: Understand the ethical implications of AI, including data privacy, bias, and transparency, and learn how to implement AI responsibly.

4. Increased Efficiency: Discover AI tools that streamline routine and administrative tasks, freeing you to focus on teaching and meaningful student interaction.

5. Personalized Learning: Learn how to use AI to create personalized learning interactions that cater to unique student needs, improving engagement and outcomes.

6. Professional Growth: Explore AI-powered professional development opportunities, enabling continuous learning and peer collaboration.

7. Future Readiness: Stay updated with the latest trends and future directions in AI, preparing yourself and your students for a rapidly evolving educational landscape.

Expected Outcomes for the Reader

By the end of this book, you will be equipped to:

- **Confidently Use AI Tools:** Integrate AI tools into your classroom confidently, leveraging them to enhance teaching and learning.

- **Personalize Learning Experiences:** Create personalized learning pathways for your students, addressing diverse needs and improving engagement.

- **Streamline Administrative Tasks:** Use AI to automate routine tasks, freeing up time for more constructive interactions with students.

- **Foster Ethical AI Practices:** Implement AI responsibly, ensuring fairness, transparency, and accountability in its use.

- **Stay Updated and Innovative:** Continuously learn and adapt to new AI technologies and methodologies, staying ahead of the curve in educational innovation.

Addressing the Reader's Pain Points

We understand that integrating AI into your teaching practices can be challenging. This book is designed to address common pain points and provide practical solutions.

Pain Point: Lack of Understanding and Confidence

Solution: This book starts with the basics, breaking down complex AI terminology into simple, understandable terms. Each chapter builds on the previous one, progressively deepening your understanding of and confidence in using AI.

Pain Point: Time Constraints

Solution: We provide practical tips and tools to streamline administrative tasks using AI, freeing up more time for teaching and interacting with students. Additionally, step-by-step guides and real-world examples make it easy to implement AI without overwhelming your schedule.

Pain Point: Ethical Concerns

Solution: This book emphasizes ethical considerations. We explore data privacy, bias, and transparency and offer guidelines for responsible AI use. You will learn how to implement AI ethically, ensuring it benefits all students somewhat.

Pain Point: Diverse Student Needs

Solution: AI offers powerful tools for personalizing learning experiences. This book provides strategies and examples of using AI to address diverse student needs, from struggling learners to advanced students, ensuring that every student can thrive.

Pain Point: Keeping Up with Technology

Solution: We offer resources and strategies for continuous learning and professional development in AI. You can remain at the forefront of educational innovation by keeping informed of the latest developments and tools.

Call to Action

As you embark on this journey, we encourage you to approach AI with an open mind and a willingness to explore its possibilities. View this book as a tool for professional growth and innovation, and use the knowledge and strategies provided to transform your teaching practices.

Take small steps—start by integrating one AI tool into your classroom, observe its impact, and build from there. Share your experiences with colleagues, collaborate, and learn together. Remember, the goal is not to replace human ingenuity with technology but to enhance it, creating a more engaging, personalized, and efficient learning environment for your students.

Concluding Thoughts

The integration of AI in education is an exciting and transformative journey. By embracing AI, you can unlock new possibilities in teaching and learning, making education more personalized, inclusive, and effective. This book is your guide to navigating this journey, providing you with the knowledge, practical skills, and insights necessary to harness the power of AI in your classroom.

We can reimagine education, creating environments where teachers and students can thrive. We invite you to join us on this journey, exploring AI's transformative potential in education and becoming a pioneer in the future of learning.

2

Notes for the Reader

AI and Regulation

The use of AI is not just increasing; it's skyrocketing. Yet, it remains largely unregulated, which is a cause for concern. To prevent potential misuse and ensure AI's ethical and responsible use, it is crucial to establish regulatory mechanisms before AI is widely implemented. Context-specific regulatory frameworks should be designed and enforced.

US-based Examples

Most AI examples cited in this book are from the US, as much of the AI regulatory development has occurred within the US context. However, as other countries adopt AI and their examples become widely cited, it is crucial that future writings include these diverse international experiences to ensure a comprehensive and global perspective on AI regulation.

Repetition of AI Applications and Examples in the Book

The varied applications of AI tools are repeatedly cited across various chapters to demonstrate their diversity of use. These illustrations help readers understand the versatility and practicality of AI tools and applications and inspire optimism about AI's potential in diverse fields.

3

Chapter 1: Demystifying AI for Educators

1.1 Decoding AI: A Glossary for Educators

A s we embark on the journey to integrate AI into education, it is essential to understand the terminology. The language of AI can seem daunting, but breaking down complex terms into more straightforward definitions can help build a strong foundation. This understanding will directly translate into practical applications in the classroom, making the learning experience more engaging and effective. For instance, AI can provide tailored learning experiences, adapt to unique student needs, and offer real-time feedback, enhancing student engagement and learning outcomes.

Key AI Terms and Definitions

1. Artificial Intelligence (AI): AI is a technology simulation of human intelligence, which thinks and learns like humans. Examples include speech recognition, problem-solving, and learning. Other examples of AI technologies include computer vision, which enables machines to 'see' and interpret visual data, and expert systems, which use AI to mimic the decision-making process of a human expert in a specific domain.

2. Machine Learning (ML): A subset of AI, ML involves training algorithms to learn from and make predictions based on data. For example, an AI system can learn to recognize patterns in student performance to provide personalized feedback.

3. Neural Networks: These algorithms modeled after the human brain recognize patterns and interpret data through interconnected nodes or "neurons." Neural networks power applications like image and speech recognition.

4. Natural Language Processing (NLP): NLP allows machines to understand and respond to human language. An example is a chatbot that can answer student queries or provide tutoring in real-time.

5. Deep Learning: An element of ML, deep learning applies neural networks with many layers (thus the term "deep") to analyze complex data. For example, deep learning can help develop sophisticated language translation tools.

6. Algorithm: A set of rules or instructions given to an AI system to help it learn and make decisions. In education, algorithms can be used to grade essays or detect plagiarism.

7. **Big Data:** Refers to analyzing massive datasets to reveal and inform patterns, trends, and associations. Big data in education can help track student performance and identify areas needing intervention.

8. **Predictive Analytics:** The use of historical data to predict future outcomes. In education, predictive analytics can forecast which students are struggling and require additional support.

9. **Robotics:** The design, construction, and use of robots to perform tasks. Educational robots can teach coding and STEM concepts through interactive activities.

10. **Chatbot:** A computer application that simulates internet-based conversation with human users. In education, chatbots can assist with administrative tasks and student inquiries.

Contextualizing AI Terms

Understanding these terms in context can make them more relatable. Let's consider a real-world scenario:

Imagine a classroom where students use tablets to complete assignments. The educator uses a learning management system (LMS) equipped with AI to track student progress. Machine learning uses algorithms to analyze student data, providing insights into their strengths and weaknesses. For example, if a student struggles with math problems related to fractions, the AI can suggest additional resources or tailored exercises. In a different setting, AI can be used in a university lecture hall to analyze student engagement and adjust the pace of the lecture accordingly.

Furthermore, the system incorporates **natural language processing** (NLP) to assist students in real-time. When a student types a question into the chatbot, NLP enables the chatbot to understand the query and respond appropriately, such as explaining a concept or directing the student to relevant materials.

1.2 The Evolution of AI in Educational Contexts

Significant milestones have marked AI's journey from concept to classroom. Understanding this evolution provides a historical perspective and inspires policymakers and technologists to appreciate the potential and limitations of current AI applications, fostering a more informed and forward-thinking approach to AI integration in education.

Tracing AI's Origins

AI has its roots in the early 20th century, but John McCarthy did not coin the term "artificial intelligence" until the 1950s. Early AI research focused on problem-solving and symbolic methods. In the 1980s, the introduction of machine learning marked a shift towards data-driven approaches, allowing systems to improve through experience.

The advent of big data and advancements in computing power in the 2000s propelled AI into mainstream applications, including education. Early educational AI systems were limited to basic computer-assisted instruction and rudimentary tutoring programs. However, the field quickly advanced with the introduction of intelligent tutoring systems in the 1980s and the rise of massive open online courses (MOOCs) in the 2010s. Today, AI is more sophisticated and capable of personalizing learning experiences and providing real-time feedback.

Highlighting Landmark Developments

Several key developments have shaped AI in education:

- **PLATO (1960s):** One of the earliest computer-based education systems, PLATO offered lessons and assessments in various subjects. While not AI-driven, it laid the groundwork for future educational technologies.

- **Intelligent Tutoring Systems (ITS) (1980s):** These systems used AI to provide personalized instruction and feedback. For example, the Carnegie Learning software developed by the Pittsburgh Advanced Cognitive Tutor Center provided customized math tutoring based on cognitive science research.

- **Massive Open Online Courses (MOOCs) (2010s):** Platforms like Coursera and edX use AI to grade assignments, provide personalized course recommendations, and enhance student engagement through interactive content.

- **AI-Powered Adaptive Learning Platforms (2010s-Present):** Systems like DreamBox and Knewton adjust content in real-time based on student performance, offering a tailored learning experience.

Understanding AI's Impact on Teaching Methods

AI has transformed traditional teaching methodologies in several ways:

- **Personalized Learning:** AI can tailor educational content to meet individual student needs, ensuring that each student progresses at their own pace. For instance, Khan Academy uses AI to recommend practice exercises based on student performance.

- **Automated Grading:** AI systems can grade assignments, providing immediate feedback while saving teachers time and helping students learn from their mistakes quickly.

- **Data-Driven Insights:** AI analyzes large datasets to provide insights into student performance, helping educators identify at-risk students and intervene early. For example, early warning systems in higher education can predict student dropout rates and suggest interventions.

- **Interactive and Engaging Content:** AI can create interactive simulations and virtual environments, making learning more engaging. Tools like zSpace offer virtual reality experiences for subjects like biology and physics.

Predicting Future Trends

The future of AI in education is promising, with several trends expected to shape the landscape:

- **Enhanced Personalization:** AI will continue to improve its delivery of highly personalized learning experiences while adapting to academic needs and emotional and social cues.

- **Increased Collaboration:** AI will facilitate greater collaboration among students and between students and teachers through intelligent collaborative tools.

- **Lifelong Learning:** AI will support lifelong learning by providing tailored educational opportunities beyond formal schooling, helping individuals continuously upgrade their skills.

As AI becomes more prevalent, ethical considerations will become increasingly important. Ensuring fairness, transparency, and accountability in AI applications, especially in education, is a shared responsibility. This commitment to the responsible use of AI should be a guiding principle for technologists and the broader community involved in education, instilling a sense of responsibility and commitment.

1.3 AI Technologies Explained Through Classroom Scenarios

To truly understand AI's potential, let's explore how various AI technologies can be applied in real-world classroom scenarios.

Introducing AI Technologies

1. **Learning Management Systems (LMS):** Platforms like Moodle and Blackboard are evolving with AI to provide personalized learning paths and automate administrative tasks.

2. **Adaptive Learning Technologies:** These systems adjust the content based on real-time analysis of student performance. DreamBox and Knewton are examples that offer personalized math and reading instruction.

3. **Intelligent Tutoring Systems (ITS):** AI-powered tutors like Carnegie Learning and Squirrel AI provide one-on-one tutoring, mimicking the personalized attention of a human tutor.

4. **Chatbots:** Tools like Ivy.ai and AdmitHub answer student queries, provide information, and assist with administrative tasks.

5. Predictive Analytics: Platforms like BrightBytes and Early Warning Systems (EWS) analyze data to predict student outcomes and suggest interventions.

6. Virtual Reality (VR) and Augmented Reality (AR): Tools like zSpace and Google Expeditions offer immersive learning experiences, making complex subjects like anatomy and history more accessible.

Applying AI to Real-World Scenarios

Scenario 1: Personalized Math Instruction

Imagine a middle school math class where each student logs into an adaptive learning platform like DreamBox. The AI system analyzes their performance on previous exercises and adapts the difficulty level of new problems accordingly. For example, if a student struggles with multiplication, the system provides additional practice and instructional videos until the student masters the concept.

Scenario 2: Real-Time Feedback on Writing Assignments

In a high school English class, students submit essays through an LMS integrated with an AI grading system like Grammarly. The AI provides immediate feedback on grammar, style, and coherence, allowing students to make revisions before final submission. This process improves writing skills and reduces the grading burden on teachers.

Scenario 3: Virtual Science Labs

A biology teacher uses VR headsets from zSpace to conduct virtual dissections. Students explore the anatomy of a frog without the ethical and logistical challenges of traditional dissections. The AI guides them through the process, providing detailed explanations and interactive quizzes to reinforce learning.

Solving Educational Challenges with AI

Challenge: Identifying At-Risk Students

A high school uses a predictive analytics platform like BrightBytes to analyze student data, including attendance, grades, and behavioral records. The AI identifies students at risk of dropping out and suggests interventions, such as counseling or tutoring. This proactive approach helps educators provide targeted support before issues escalate.

Challenge: Enhancing Student Engagement

An elementary school teacher uses AI-powered gamification tools like Kahoot! to create interactive quizzes. The AI adapts the questions based on student responses, keeping them engaged and motivated. Additionally, the system tracks student progress, providing valuable insights for personalized instruction.

Challenge: Providing Accessible Education

A university implements an AI-driven lecture transcription service, making content accessible to students with hearing impairments. The AI also translates the transcripts into multiple languages, supporting international students. This inclusive approach enables students to have equal access to educational resources.

Encouraging Practical Application

Educators can start integrating AI into their classrooms by:

- **Exploring AI Tools:** Experiment with free or low-cost AI tools available online to understand their functionalities and potential benefits.

- **Collaborating with Colleagues:** Share experiences and insights with fellow educators to learn from each other's successes and challenges.

- **Seeking Professional Development:** Participate in workshops, webinars, and courses on AI in education to stay updated with the latest trends and best practices.

1.4 Addressing Concerns and Ethical Considerations

Integrating AI into education raises important ethical and practical considerations, as with any transformative technology. Addressing these concerns promptly and comprehensively ensures that AI is used responsibly and effectively.

Data Privacy and Security

Concern: Protecting student data is paramount. AI collects and analyzes massive volumes of data, raising concerns about privacy and security.

Solution: Implement robust data protection policies and use secure, encrypted systems. Educators should comply with AI tool regulations, including the General Data Protection Regulation (GDPR) and the Family Educational Rights and Privacy Act (FERPA).

Bias and Fairness

Concern: AI systems can perpetuate and exacerbate data biases, resulting in unfair outcomes, particularly for marginalized groups.

Solution: Train AI systems using diverse and representative datasets. Regularly audit AI algorithms for bias and implement mechanisms to address and rectify any identified biases.

Transparency and Accountability

Concern: AI systems' decision-making processes can be opaque, making it difficult to understand how conclusions are reached.

Solution: Promote transparency by using explainable AI models that provide clear insights into how decisions are made. Establish accountability frameworks to ensure that AI applications are used ethically and responsibly.

Teacher and Student Roles

Concern: Some fear that AI might replace teachers or reduce human interaction in the classroom.

Solution: Emphasize that AI is a tool to augment, not replace, human educators. AI can handle repetitive tasks, allowing teachers to focus on more meaningful student interactions. Encourage a balanced approach where technology complements human expertise.

Equity and Access

Concern: There is a risk that AI could widen the digital divide, limiting access to advanced technologies for disadvantaged students.

Solution: Advocate for equitable access to AI tools and resources. Implement initiatives to provide necessary technology and training to underserved communities, ensuring all students benefit from AI advancements.

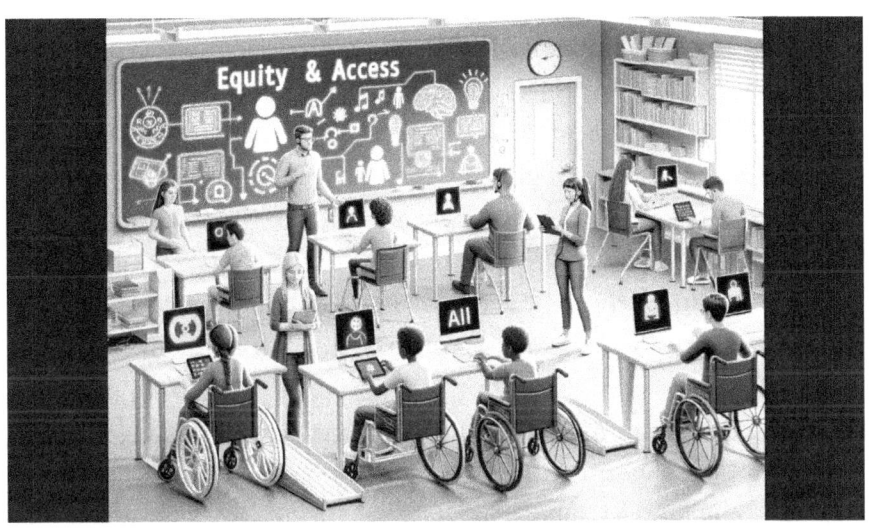

1.5 The Road Ahead: Building a Vision for AI-Enhanced Education

The potential of AI in education is vast, but realizing its complete potential calls for a collective effort on the part of educators, policymakers, technologists, and the broader community. By embracing AI thoughtfully and ethically, we can create a future where education is more personalized, inclusive, and effective.

Envisioning the Future Classroom

In the AI-enhanced classroom of the future:

- **Personalized Learning:** Each student has a personalized learning path, with AI continually adapting content to meet their needs and preferences.

- **Collaborative Learning:** AI facilitates collaboration through intelligent tools supporting group projects and peer learning, fostering community.

- **Real-Time Feedback:** AI provides instant feedback on assignments, enabling students to learn from past performance and improve continuously.

- **Inclusive Education:** AI ensures that all students have access to high-quality education and support regardless of their backgrounds or abilities.

- **Lifelong Learning:** AI supports lifelong learning by providing personalized educational opportunities beyond formal schooling, helping individuals adapt to a rapidly changing world.

Actionable Steps for Educators

To move towards this vision, educators can:

- **Stay Informed:** Use professional development and networking with peers to remain up-to-date with the latest developments in AI and education.

- **Experiment and Innovate:** Start small by integrating AI tools into specific aspects of teaching and gradually expand based on what works best.

- **Collaborate and Share:** Work with colleagues, tech developers, and policymakers to share insights, address challenges, and co-create effective AI solutions.

- **Advocate for Equity:** Champion initiatives that ensure equitable access to AI technologies and resources for all students.

Concluding Thoughts

Integrating AI in education involves adopting new technologies and reimagining how we teach and learn. By embracing AI with an open mind and a commitment to ethical and equitable practices, we can transform education for the better, preparing students for a future where they can thrive.

This chapter examines the role of AI in education in depth. As we delve into specific applications, case studies, and practical strategies in the subsequent chapters, remember that the goal is to enhance the human aspects of education—fostering creativity, critical thinking, and a love for learning. AI is a powerful tool to help achieve these goals, but educators' passion and dedication will ultimately drive the transformation.

Chapter 2: Ethical Considerations in AI

2.1 Data Privacy 101 for Educators Using AI

Understanding Data Privacy

I ncorporating AI into educational settings, such as using AI-powered adaptive learning platforms or AI-driven student performance analytics, has immense potential to enhance learning experiences and streamline administrative tasks. However, it also raises critical concerns about data privacy. As educators, your understanding and proactive approach to these concerns are essential and empowering. Your role in protecting students' personal information and building trust with all stakeholders is crucial and should be acknowledged.

Why Data Privacy Matters

Data privacy refers to the handling, storing, and protecting of personal information to ensure it is not accessed or misused by unauthorized parties. In the context of AI in education, data privacy is paramount because:

- **Sensitive Information:** Schools collect vast amounts of sensitive information, including student performance data, health records, and personal identifiers.

- **Student Safety:** Protecting this data ensures the safety and well-being of students, preventing misuse that could lead to identity theft, cyberbullying, or other serious harms. For instance, if student data is not adequately protected, it could be used to impersonate a student, leading to potential identity theft or cyberbullying incidents.

- **Legal Compliance:** Schools must comply with legal data protection laws and regulations, including the Family Educational Rights and Privacy Act (FERPA) and the General Data Protection Regulation (GDPR).

Best Practices for Data Protection

Implementing best practices for data protection involves several steps:

1. Data Minimization: Collect only the data necessary for educational purposes. Avoid gathering excessive or unrelated information that could increase the risk of breaches.

2. Encryption: Encryption protects data at rest and in transit. Even if data is intercepted, it cannot be read without the encryption key.

3. Access Controls: Ensure strict access controls permitting only authorized personnel to access sensitive data. Use multi-factor authentication and role-based access to enhance security.

4. Regular Audits: Conduct audits to identify potential vulnerabilities and ensure compliance with data protection policies. This robust strategy helps maintain a secure data environment, giving you confidence that your efforts are practical.

5. Training and Awareness: Educate and sensitize staff and students about data privacy and best practices for protecting personal information. As educators, you play a crucial role in this process. By understanding the data privacy risks associated with AI and implementing the best practices we've discussed, you can help ensure the security of student data. Your role may include educating students about the importance of keeping their login credentials or personal information confidential and ensuring that the AI tools you use in your classroom comply with data protection laws. However, it's important to note that ensuring data privacy is a collective effort, and other stakeholders, such as school administrators and policymakers, also have important roles to play.

Legal Considerations

Compliance with legal frameworks is crucial for protecting student data:

- **Family Educational Rights and Privacy Act (FERPA):** The FERPA is a US law that protects the privacy of education records. It gives parents certain rights concerning their children's education records, such as controlling who can access them. Schools must obtain consent in writing from parents, guardians, or eligible students before disclosing personally identifiable information stored in education records. This is particularly relevant when using AI tools that collect or use student data. For instance, if you're using an AI-driven student performance analytics system, you must ensure that you have the consent from parents or eligible students to collect and use their academic data. The General Data Protection Regulation (GDPR) is a similar law in the European Union that mandates strict requirements for data protection, including obtaining explicit permission for the collection and the right to access and delete data.

- **General Data Protection Regulation (GDPR):** The GDPR mandates strict requirements for data protection in the European Union, including obtaining explicit consent for data collection and the right to access and delete individual data.

Building Trust with Stakeholders

Transparent communication with parents, students, and the community about data privacy practices is essential:

- **Privacy Policies:** Develop clear privacy policies outlining how data is collected, used, and protected. Make these policies easily accessible to all stakeholders.

- **Parental Consent:** Obtain informed consent from parents before collecting or using student data for AI applications. Explain the benefits and risks associated with data usage.

- **Regular Updates:** Provide regular updates on data protection measures and any changes to privacy policies. This is important for keeping stakeholders informed and reassured about the security of their data. Regular updates can include information about new data protection measures, changes in privacy policies, or updates on ongoing data protection efforts. This regular communication helps build trust and maintain transparency with stakeholders.

2.2 Ethical AI Deployment: Best Practices

Developing Ethical AI Usage Policies

Creating and implementing ethical AI usage policies ensures that AI tools are used responsibly and transparently. These policies should address several key areas:

1. Purpose and Scope: Clearly define the purpose of using AI in the educational context. Specify the types of AI applications being deployed and their intended outcomes.

2. Transparency: Ensure that the AI systems used are transparent. Educators, students, and parents must know how AI algorithms make decisions and what data they use.

3. Consent: Obtain explicit consent from students and parents for using AI tools, especially when collecting personal data. This includes informing them about the nature and purpose of the data collection, how it will be used, and who will have access to it. Your understanding and communication of these aspects are crucial. Obtaining consent is a key principle of data privacy and protection, and it's essential when using AI tools that collect or use personal data.

4. Bias and Fairness: Implement measures to identify and mitigate biases in AI algorithms. Ensure that AI applications are fair and do not disproportionately disadvantage any group of students.

5. Accountability: Establish precise accountability mechanisms for the use of AI. This includes regular audits, reviews, and the ability to address grievances related to AI usage. It's important to note that ensuring ethical AI use is not a one-time task but an ongoing process. As new AI technologies and applications emerge, staying updated and adapting your practices is crucial to ensure continued ethical use.

Inclusive AI Design Principles

Inclusive design ensures that AI tools cater to the diverse needs of all students, including those with disabilities or from different cultural backgrounds. It's about designing AI tools usable to as many people as possible, regardless of their abilities or backgrounds. This means that AI tools should be designed to be used by all students, irrespective of their learning styles, abilities, or cultural backgrounds. As educators, your insights and feedback are invaluable in this process. Key principles include:

Universal Design for Learning (UDL): AI tools should be designed according to UDL principles, providing multiple means of engagement, representation, and expression to accommodate diverse learning styles and abilities.

- **Accessibility:** Ensure that AI applications are accessible to students with disabilities. This includes screen readers, alternative input methods, and customizable interfaces.

- **Cultural Sensitivity:** Design AI tools that respect and reflect the diversity of ethnic and cultural contexts of the student community. This includes language support, culturally relevant content, and sensitivity to social norms and values. For instance, if you're using an AI-powered language learning platform, it should be able to support multiple languages, offer learning materials relevant to different cultures, and respect social norms and values in its interactions with students. Another example could be an AI tool that provides content recommendations based on students' cultural interests and preferences.

Transparency in AI Applications

Transparency in AI tools and applications is critical to ensure ethical use and trust in technology. In the context of AI, transparency means being open and transparent about how AI systems work, how they make decisions, and what data they use. This helps educators and students understand and trust the AI's recommendations and interventions. For instance, if you're using an AI-powered adaptive learning platform, you should be able to explain to your students how the platform makes recommendations based on their learning history and preferences. This transparency can help students understand and trust the platform's recommendations.

- **Explainability:** AI systems should be explainable, meaning non-experts can comprehend their decision-making strategies. This is key to building trust in AI. For instance, if an AI system recommends a particular learning resource for a student, it should be able to explain why that resource was chosen based on the student's learning history and preferences. This helps educators and students trust the AI's recommendations and interventions. Explainability is particularly important when using AI tools in education, as it allows educators and students to understand and trust the AI's decisions.

- **Documentation:** Maintain comprehensive documentation of AI systems, including their design, data sources, and decision-making criteria. This documentation should be accessible to all stakeholders.

- **User Control:** Provide users with control over AI applications, including the ability to opt out or adjust AI settings according to their preferences and needs.

Accountability Mechanisms

Ensuring accountability involves establishing checks and balances to monitor and regulate AI usage:

- **Regular Audits:** Conduct audits of AI systems to identify and address ethical issues, biases, or unintended consequences.

- **Ethics Committees:** Establish ethics committees to oversee the deployment and use of AI in educational settings. These committees can provide guidance, address concerns, and ensure compliance with ethical standards.

- **Feedback Mechanisms:** Implement feedback mechanisms that allow students, parents, and educators to report issues or concerns about AI usage. Use this feedback to improve and refine AI applications.

2.3 Navigating Bias and Fairness in AI for Classroom Use

Identifying Bias in AI

Bias in AI can arise from various sources, including biased training data, biased algorithm design, or biased deployment practices. Identifying and addressing these biases is crucial for fair and equitable AI usage. In education, bias in AI can lead to unfair grading, unequal access to learning resources, or reinforcement of existing inequalities. Educators need to be aware of these risks and take steps to mitigate them.

Sources of Bias

- **Data Bias:** If the training data used to develop AI models is biased, the AI's decisions will reflect those biases. For example, if an AI system is trained on data that predominantly represents one demographic group, it may perform poorly for other groups. This could lead to biased grading, unequal access to learning resources, or reinforcing existing inequalities.

- **Algorithmic Bias:** Biases can also be introduced to AI systems and tools during the algorithm design phase. This can occur if the algorithm's objectives or constraints unintentionally favor certain groups over others.

- **Deployment Bias:** Biases can emerge during deployment if the AI system is not adequately tested and validated for diverse student populations.

Strategies for Identifying Bias

- **Diverse Datasets:** Use diverse and representative datasets to train AI models. This helps ensure that the AI system performs well across different student groups.

- **Bias Audits:** Conduct regular bias audits to identify and address biases in AI algorithms. This involves testing the AI system on various demographic groups and analyzing its performance.

- **Stakeholder Input:** Involve diverse stakeholders, including students, parents, and educators, in the AI development process. Their input can help identify potential biases and improve the AI system's fairness.

Mitigating Bias

Once biases are identified, steps must be taken to mitigate them:

- **Data Preprocessing:** Use data preprocessing techniques to address biases in the training data. This can include balancing the dataset, removing biased attributes, or augmenting the data with underrepresented groups.

- **Algorithmic Adjustments:** Adjust the AI algorithms to minimize bias. This can involve reweighting the training data, adding fairness constraints to the algorithm, or using bias-correction techniques.

Continuous Monitoring: After deployment, continuously monitor the AI system for biases. This involves regularly testing the system, analyzing its performance, and adjusting to maintain fairness.

Promoting Fairness

Promoting fairness involves ensuring that AI tools are used equitably across all student populations:

- **Equal Access:** Ensure all students access AI tools and resources equally. This includes providing necessary technology and support to underserved communities.

- **Fair Evaluation:** Use AI tools to evaluate students fairly, without favoritism or discrimination. This includes ensuring that automated grading systems are unbiased and provide accurate assessments.

Inclusive Practices: Implement inclusive practices that consider students' diverse needs. This includes using AI tools to provide personalized learning experiences tailored to individual strengths and challenges.

Case Studies: Addressing Bias and Promoting Fairness

Case Study 1: Mitigating Bias in Predictive Analytics

A high school implemented an AI-based early warning program to identify those at risk of dropping out. Initial results showed that the system disproportionately flagged minority students. The school addressed this bias by:

- **Using Diverse Data:** Incorporating diverse data sources that better represent the student population.

- **Bias Audits:** Conducting regular bias audits to identify and correct biases in the algorithm.

- **Stakeholder Engagement:** Engaging with minority students, parents, and community leaders to gather feedback and improve the system's fairness.

Case Study 2: Ensuring Fairness in Automated Grading

A university adopted an AI-powered essay grading system to streamline assessments. However, students reported that the system was biased against non-native English speakers. The university addressed this issue by:

- **Bias Correction:** Adjusting the algorithm to account for linguistic diversity and minimize bias against non-native speakers.

- **Human Review:** Implementing a system where human graders reviewed AI-graded essays to ensure fairness and accuracy.

- **Student Feedback:** Collecting and incorporating student feedback to continuously improve the grading system.

2.4 Case Studies: Ethical Dilemmas and Solutions in AI Education

Analyzing Ethical Dilemmas

Exploring real-world ethical dilemmas provides valuable insights into the challenges and solutions to using AI in education. Here are some illustrative case studies:

Case Study 1: Balancing Privacy and Learning Analytics

Dilemma: A school district implemented an AI-based learning analytics system to track student performance and provide personalized feedback. However, parents raised concerns about the extensive data collection and potential privacy breaches.

Solution:

- **Transparent Communication:** The school district held informational sessions with parents to explain how the data was collected, used, and protected. They provided detailed privacy policies and reassured parents that data was anonymized.

- **Opt-In/Opt-Out Options:** The school respected parents' preferences and concerns by offering them the option to opt in or out of the analytics program.

- **Data Minimization:** The district minimized data collection to essential information needed for educational purposes, reducing the risk of privacy breaches.

Outcome: By addressing privacy concerns transparently and offering choices, the school district gained parental trust and successfully implemented the learning analytics system.

Case Study 2: Addressing Algorithmic Bias in Admissions

Dilemma: A university used an AI system to streamline the admissions process. However, the system favored applicants from specific socioeconomic backgrounds, raising concerns about fairness and equity.

Solution:

- **Bias Detection:** The university conducted a thorough analysis of the AI system to identify and understand the biases.

- **Algorithm Adjustments:** The algorithm was adjusted to consider a broader range of factors and did not disproportionately favor any group.

- **Diverse Data:** The university included more diverse data in the training set to improve the system's fairness.

Outcome: The university improved the fairness of its admissions process, ensuring that all applicants were evaluated equitably, regardless of their socioeconomic background.

Solutions and Outcomes

Addressing ethical dilemmas requires thoughtful solutions and a commitment to continuous improvement:

- **Proactive Measures:** To prevent ethical issues, implement proactive measures such as transparency, inclusivity, and stakeholder engagement.

- **Responsive Actions:** When ethical dilemmas arise, respond promptly and transparently, involving all relevant stakeholders in the solution process.

- **Continuous Improvement:** Monitor and refine AI systems to ensure they remain ethical, fair, and effective.

Expert Insights

Incorporating expert insights provides depth and perspective to the discussion on ethical AI:

- **AI Ethics Experts:** Consult with AI ethics experts to understand the broader ethical implications and best practices for AI deployment.

- **Interdisciplinary Collaboration:** Develop comprehensive ethical guidelines by collaborating with experts from various fields, including education, technology, ethics, and law.

- **Community Involvement:** Engage with the broader community, including parents, students, and educators, to gather diverse perspectives and build consensus on ethical AI practices.

Guidelines for Ethical Decision-Making

To support educators in making ethical decisions when implementing AI, here are some practical guidelines:

1. Define Clear Objectives: Define the objectives of using AI in education, ensuring they align with educational goals and ethical standards.

2. Involve Stakeholders: Involve all relevant stakeholders, including students, parents, educators, and community members, in the decision-making process.

3. Prioritize Transparency: Ensure transparency in how AI systems operate, what data they use, and how decisions are made.

4. Conduct Regular Audits: Regularly audit AI systems for biases, ethical issues, and unintended consequences. Use these audits to make necessary adjustments.

5. Foster Inclusivity: Design AI tools that are inclusive and accessible to all students, considering diverse needs and backgrounds.

6. Educate and Train: Provide education and training on ethical AI practices for educators, students, and parents to build awareness and understanding.

7. Establish Accountability: Create accountability mechanisms to monitor AI usage and promptly address ethical concerns.

Conclusion

Ethical considerations are fundamental to the successful integration of AI in education. Educators can use AI tools responsibly and effectively by understanding and addressing data privacy, bias, transparency, and accountability. Through real-world examples and expert insights, we have explored practical strategies to ensure ethical AI deployment. By following these guidelines, educators can unlock AI's potential to enhance outcomes in teaching and learning while upholding the highest ethical standards.

The next chapter will provide practical strategies for integrating AI into educational practices, focusing on creating AI-enhanced lesson plans, personalizing learning pathways, and simplifying administrative tasks as we move forward in this book. The journey to ethical and practical AI in education continues, empowering educators to innovate and inspire.

5

Chapter 3: AI Integration Strategies

3.1 Creating AI-Enhanced Lesson Plans

I ntegrating AI into lesson plans is a transformative approach that can significantly improve the learning experience. By making it more engaging, personalized, and effective, AI tools are becoming increasingly essential in educational settings. This section provides a comprehensive guide to carving AI-enhanced lesson plans, ensuring alignment with curriculum goals, and demonstrating the potential benefits through specific examples.

Step-by-Step Guide to Incorporating AI Tools

Step 1: Identify Learning Objectives

Before integrating AI, clearly define the learning objectives for your lesson. What skills or knowledge do you want students to acquire? For example, the aim of a high school math class might be to master quadratic equations.

Step 2: Select Appropriate AI Tools

When selecting AI tools, consider those that align with your learning objectives. For instance, in a math lesson on quadratic equations, you could leverage the power of DreamBox, an adaptive learning platform that generates personalized math problems tailored to each student's performance.

Step 3: Plan the Lesson Structure

Structure your lesson to incorporate AI tools effectively. Break down the lesson into segments: introduction, guided practice, independent practice, and assessment. Determine where AI tools will be most beneficial. For example:

- **Introduction:** Use an AI-powered interactive video to introduce the concept of quadratic equations.

- **Guided Practice:** Employ an AI tutor like Carnegie Learning to provide step-by-step guidance through sample problems.

- **Independent Practice:** Use DreamBox to assign personalized exercises based on each student's proficiency level.

- **Assessment:** Use an AI grading tool like Gradescope to grade student submissions and automatically provide immediate feedback.

Step 4: Integrate AI into Lesson Activities

Incorporate AI tools into various activities within your lesson. For example:

- **Interactive Videos:** Use platforms like Edpuzzle to create interactive videos with embedded quizzes that adapt based on student responses.

- **AI Tutors:** Implement AI tutors that offer personalized hints and explanations as students work through problems.

- **Adaptive Learning Platforms:** Assign tasks on adaptive learning platforms that adjust difficulty levels in real-time.

- **AI Grading Tools:** Use AI tools to grade assignments instantly, allowing students to learn from their mistakes and improve.

Step 5: Monitor and Adjust

Throughout the lesson, monitor student progress and engagement. Use data from AI tools to adjust your teaching strategies as needed. For instance, if DreamBox data shows that several students struggle with a specific type of quadratic equation, you can revisit that topic in class.

Step 6: Reflect and Iterate

After the lesson, reflect on what worked well and could be improved. Gather feedback from students and analyze data from AI tools to refine your lesson plans for future use.

Aligning AI with Curriculum Goals

Ensuring that AI tools help achieve specific learning objectives is crucial. Here are examples of how AI can be aligned with curriculum goals in various subjects:

Mathematics

- **Objective:** Improve problem-solving skills in algebra.

- **AI Tool:** Khan Academy's AI-driven platform provides personalized practice problems and instant feedback.

- **Outcome:** Students receive tailored problem sets that adapt to their learning pace, helping them master algebraic concepts more effectively.

Science

- **Objective:** Understand the scientific method through hands-on experiments.

- **AI Tool:** Labster's virtual labs offer interactive simulations of scientific experiments.

- **Outcome:** Students can conduct virtual experiments, making the scientific method tangible and engaging without the constraints of a physical lab.

Language Arts

- **Objective:** Enhance writing skills and comprehension.

- **AI Tool:** Grammarly provides real-time feedback for grammar, style, and coherence.

- **Outcome:** Students improve their writing skills through immediate, personalized feedback, allowing them to refine their drafts iteratively.

History

- **Objective:** Analyze historical events and their impacts.

- **AI Tool:** Google Arts & Culture offers virtual tours and interactive timelines.

- **Outcome:** Students explore historical events through immersive experiences, deepening their understanding and engagement with the material.

Example Lesson Plans with AI Integration

Example 1: High School Algebra Lesson

Learning Objective: Master solving quadratic equations

Lesson Plan

1. Introduction (10 minutes):

- Use an AI-powered interactive video on Edpuzzle to introduce quadratic equations.

- Embed quizzes to check for understanding and adjust the content based on student responses.

2. Guided Practice (20 minutes):

- Utilize Carnegie Learning's AI tutor to walk students through solving quadratic equations step-by-step.

- The AI provides hints and explanations tailored to each student's progress.

3. Independent Practice (30 minutes):

- Assign personalized exercises on DreamBox. The platform adjusts the difficulty of problems based on student performance.

- Monitor student progress through the platform's analytics dashboard.

4. Assessment (15 minutes):

- Use Gradescope to grade student submissions automatically. Provide instant feedback highlighting areas for improvement.

5. Reflection (10 minutes):

- Discuss common challenges and misconceptions identified through AI analytics.

- Encourage students to reflect on their learning and set goals for improvement.

Example 2: Middle School Science Lesson

Learning Objective: Understand the principles of chemical reactions

Lesson Plan

1. Introduction (15 minutes):

- Start with an AI-driven interactive video from Edpuzzle explaining chemical reactions.

- Use embedded quizzes to assess prior knowledge and adjust content accordingly.

2. Guided Practice (25 minutes):

- Conduct a virtual lab experiment on Labster where students simulate a chemical reaction.

- The AI guides them through the process, providing real-time feedback and explanations.

3. Independent Practice (30 minutes):

- Assign additional virtual experiments on Labster. The platform adapts to each student's progress and provides individualized feedback.

4. Assessment (20 minutes):

- Use an AI grading tool to evaluate lab reports. Provide instant feedback on accuracy and understanding.

5. Reflection (10 minutes):

- Discuss the experiment results and insights gained from AI feedback.

- Encourage students to think about how they can apply their knowledge to real-world scenarios.

3.2 AI Tools for Personalized Learning Pathways

AI can potentially transform personalized learning by targeting educational experiences to unique student needs. This section explores AI tools that support personalized learning pathways, offering practical examples and success stories.

Overview of Personalized Learning

Personalized learning involves creating customized educational experiences that cater to each student's unique needs, strengths, and interests. AI can enhance personalized learning by:

- **Adapting Content:** AI systems can adjust the difficulty and type of content based on real-time analysis of student performance.

- **Providing Instant Feedback:** AI tools offer immediate feedback, helping students understand their mistakes and improve.

- **Tracking Progress:** AI-powered platforms track student progress over time, identifying areas where additional support is needed.

AI Tools That Support Personalization

DreamBox Learning

- **Description:** An adaptive math program that provides personalized lessons based on student performance.

- **Application:** Students complete math exercises, and the AI adjusts the difficulty level in real-time to ensure optimal learning.

- **Outcome:** Students receive tailored instruction that meets their needs, leading to improved math proficiency.

Knewton Alta

- **Description:** An adaptive learning platform that offers personalized coursework in various subjects.

- **Application:** The platform uses AI to analyze student performance and adapt the curriculum accordingly.

- **Outcome:** Students engage with appropriately challenging material, enhancing their understanding and retention.

Carnegie Learning

- **Description:** An AI-powered tutoring system that provides personalized math instruction.

- **Application:** The AI tutor offers step-by-step guidance and adjusts the level of support based on student needs.

- **Outcome:** Students receive individualized support, improving their problem-solving skills and confidence in math.

Content Technologies, Inc. (CTI)

- **Description:** An AI company that creates personalized textbooks tailored to individual student needs.

- **Application:** The AI generates customized textbooks based on student performance data, ensuring each student receives the most relevant and helpful content.

- **Outcome:** Students benefit from learning materials that address their strengths and weaknesses.

Implementing AI for Differentiation

Step 1: Assess Student Needs

Begin by assessing the individual needs of your students. Use AI-powered diagnostic tools to gather data on their strengths, weaknesses, and learning preferences. For example, an AI diagnostic tool can identify areas where students struggle with reading comprehension.

Step 2: Choose Appropriate AI Tools

Select AI tools that align with the identified needs. For instance, if the assessment reveals that some students need additional support in reading, you might choose an adaptive reading platform like Amira Learning, which provides personalized reading practice and feedback.

Step 3: Integrate AI into Instruction

Incorporate AI tools into your instruction to provide differentiated learning experiences. For example:

- **Reading Intervention:** Use Amira Learning to provide personalized reading practice for struggling readers. The AI adjusts the difficulty of texts and offers real-time feedback on fluency and comprehension.

- **Math Differentiation:** Implement DreamBox Learning to offer tailored math exercises that adapt to each student's skill level.

- **Writing Support:** Use Grammarly to provide personalized feedback on student writing, helping them improve their grammar, style, and coherence.

Step 4: Monitor and Adjust

Continuously monitor student progress using data from AI tools. Adjust your instructional strategies based on the insights gained. For example, if the AI diagnostic tool indicates that a student has mastered a particular skill, you can provide more advanced materials to challenge them.

Step 5: Reflect and Iterate

Reflect on the effectiveness of the AI tools and make necessary adjustments. Gather feedback from students and use AI data to refine your differentiation strategies. For instance, if students report that the AI reading platform is too challenging, you can adjust the settings to provide more appropriate texts.

Success Stories

Success Story 1: Personalized Math Instruction with DreamBox Learning

Context: An elementary school implemented DreamBox Learning to support personalized math instruction.

Implementation: Teachers used DreamBox to assign math exercises tailored to each student's skill level. The AI adjusted the difficulty of problems based on student performance.

Outcome: Students showed significant improvement in math proficiency, with many advancing beyond grade-level expectations. Teachers reported that DreamBox helped them effectively identify and address individual learning needs.

Success Story 2: Customized Reading Practice with Amira Learning

Context: A middle school introduced Amira Learning to support struggling readers.

Implementation: Amira Learning provided personalized reading practice, adjusting the difficulty of texts and offering real-time feedback on fluency and comprehension.

Outcome: Students improved their reading fluency and comprehension, with many moving up several reading levels. Teachers found that Amira Learning allowed them to provide targeted support and track progress more accurately.

3.3 Simplifying Administrative Tasks with AI for Teachers

AI can significantly reduce the administrative demands on teachers, freeing their time to focus more on instruction and student interaction. This section explores AI tools that simplify administrative tasks, providing practical examples and case studies.

AI for Efficiency

Grading and Assessment

AI-powered grading tools can automate the grading process, providing instant feedback and saving teachers valuable time.

Gradescope

- **Description:** An AI-powered grading tool that automates grading assignments, quizzes, and exams.

- **Application:** Teachers upload student submissions, and the AI grades them based on predefined criteria.

- **Outcome:** Teachers save time on grading, and students receive immediate feedback on their work.

Socrative

- **Description:** An AI-driven assessment tool that provides real-time quizzes and instant feedback.

- **Application: Teachers create quizzes, which students complete** on their devices. The AI analyzes responses and provides immediate feedback.

- **Outcome:** Teachers can quickly assess student understanding and adjust instruction accordingly.

Attendance and Record-Keeping

AI tools can streamline attendance tracking and record keeping, reducing administrative workload.

Klassroom

- **Description:** An AI-powered attendance and communication tool for teachers.

- **Application:** Teachers use Klassroom to track attendance, communicate with parents, and manage classroom activities.

- **Outcome:** Attendance tracking becomes more efficient, and communication with parents is streamlined.

Automating Routine Tasks

AI can automate routine administrative tasks, freeing up more time for teaching.

Google Classroom

- **Description:** A learning management system integrating AI to streamline administrative tasks.

- **Application:** Teachers use Google Classroom to create assignments, distribute materials, and communicate with students. The AI assists with organizing and managing these tasks.

- **Outcome:** Teachers spend less time on administrative tasks and more time on instruction and student interaction.

Reducing Workload

Step 1: Identify Time-Consuming Tasks

Identify the administrative tasks that consume the most time. Typical tasks include grading, attendance tracking, and record keeping.

Step 2: Select AI Tools

Choose AI tools that can automate or streamline these tasks. For example:

- **Grading:** Use Gradescope to automate the grading process.

- **Attendance:** Implement Klassroom to streamline attendance tracking.

- **Record Keeping:** Use Google Classroom to manage assignments and student records.

Step 3: Integrate AI Tools into Daily Routine

Incorporate AI tools into your daily routine to reduce the administrative workload. For example:

- **Automated Grading:** Use Gradescope to grade assignments and provide instant feedback.

- **Streamlined Attendance:** Use Klassroom to track attendance and communicate with parents.

- **Efficient Record Keeping:** Use Google Classroom to organize and manage assignments, materials, and student records.

Step 4: Monitor and Adjust

Monitor the utility and potential of the AI tools and adjust them as needed. Gather feedback from students and use AI data to refine your administrative processes.

Step 5: Reflect and Iterate

Reflect on the impact of AI tools on your workload and teaching. Use the insights gained to improve your continuous use of AI in administrative tasks.

Case Studies

Case Study 1: Automated Grading with Gradescope

Context: A high school implemented Gradescope to automate the grading of math and science assignments.

Implementation: Teachers used Gradescope to grade assignments and provide instant feedback. The AI analyzed student submissions based on predefined criteria.

Outcome: Teachers saved significant time on grading, allowing them to focus more on teaching and meaningful student interaction. Students received immediate feedback, helping them learn from their mistakes and improve their performance.

Case Study 2: Streamlined Attendance with Klassroom

Context: An elementary school introduced Klassroom to streamline attendance tracking and communication with parents.

Implementation: Teachers used Klassroom to track attendance and interact with parents about student progress and classroom engagement.

Outcome: Attendance tracking became more efficient, and communication with parents was streamlined. Teachers reported that Klassroom saved time on routine administrative tasks, allowing them to engage more with teaching.

3.4 Strategies for Incorporating AI Without Overwhelming Your Classroom

Integrating AI into the classroom can be overwhelming if not done thoughtfully. This section provides strategies for incorporating AI in a manageable way, balancing AI tools with traditional teaching methods, and managing student expectations.

Starting Small

Begin with Simple AI Tools

Start by integrating simple AI tools that are easy to use and implement. For example:

- **Grammar and Writing:** Use Grammarly to provide instant feedback on student writing.

- **Interactive Videos:** Implement Edpuzzle to create interactive videos with embedded quizzes.

Pilot AI Tools in Specific Areas

Pilot AI tools in specific areas of your teaching to see how they work and what impact they have. For example:

- **Math Practice:** Use DreamBox Learning for personalized math practice.

- **Reading Intervention:** Implement Amira Learning for personalized reading support.

Gather Feedback and Reflect

Gather feedback from students on their experience with AI tools. Reflect on what worked well and what could be improved. Use this feedback to make informed decisions about further AI integration.

Balancing AI and Traditional Teaching

Combine AI with Traditional Methods

Balance AI tools with traditional teaching methods to create a blended learning environment. For example:

- **Lectures and Interactive Videos:** Combine traditional lectures with interactive videos from Edpuzzle.

- **Hands-On Activities and Virtual Labs:** Use virtual labs from Labster to complement hands-on science experiments.

Maintain Human Interaction

Ensure that AI tools do not replace human interaction. Use AI to enhance, not replace, teacher-student interactions. For example:

- **AI Tutors and Teacher Support:** Use AI tutors like Carnegie Learning to provide additional support while teachers focus on facilitating discussions and addressing individual student needs.

Monitor Student Engagement

Monitor student engagement with AI tools and adjust your teaching strategies accordingly. Use data from AI tools to identify areas where students need more support or challenge.

Managing Student Expectations

Set Clear Expectations

Set clear expectations for how AI tools will be used in the classroom. Explain the benefits and limitations of AI tools to students. For example:

- **Interactive Videos:** Explain that interactive videos from Edpuzzle are designed to enhance understanding and provide immediate feedback.

- **AI Grading:** Clarify that AI grading tools like Gradescope provide instant feedback, but human review is also essential for thorough assessment.

Provide Training and Support

Provide training and support to help students use AI tools effectively. For example:

- **Tutorials and Guides:** Offer tutorials and guides on how to use AI tools like DreamBox Learning or Amira Learning.

- **Ongoing Support:** Provide ongoing support and address any questions or concerns students may have about using AI tools.

Encourage a Growth Mindset

Encourage a growth mindset in students, emphasizing that AI tools support their learning and help them improve. For example:

- **Instant Feedback:** Highlight the value of instant feedback from AI tools in helping students identify and address their mistakes.

- **Personalized Learning:** Emphasize that personalized learning experiences from AI tools are designed to meet their unique needs and help them succeed.

Continuous Learning

Stay Informed About AI Trends

Stay up-to-date regarding developments and trends in AI in education. Engage with professional development opportunities and network with colleagues and peers to learn about new AI tools and best practices.

Experiment and Innovate

Experiment with different AI tools and innovative teaching strategies. Be open to trying new approaches and adapting to what works best for your students.

Collaborate with Colleagues

Collaborate with peers and colleagues to exchange insights, experiences, and lessons from AI integration in the classroom. Learn from each other and work together to create a supportive and innovative learning environment.

Conclusion

Integrating AI into the classroom offers immense potential to enhance teaching and learning, but it must be done mindfully, thoughtfully, and responsibly. By following the strategies outlined in this chapter, educators can create AI-enhanced lesson plans, personalize learning experiences, simplify administrative tasks, and incorporate AI in a manageable way. The journey to AI integration is continuous, requiring reflection, adjustment, and collaboration. As educators embrace AI, they can unlock new possibilities in education, creating engaging, personalized, and effective learning environments for all students.

Chapter 4: AI for Classroom Management and Efficiency

4.1 AI Chatbots as Teaching Assistants: A How-To Guide

A I chatbots, as practical tools, can transform classroom management by serving as teaching assistants, thereby supporting both teachers and students. This section is a practical guide that will walk you through the process of setting up and effectively using AI chatbots in your classroom.

What are AI Chatbots?

AI chatbots are computer programs designed to simulate conversations with human users. They use natural language processing (NLP) to understand and respond to queries, making them useful for various educational purposes. In the classroom, chatbots can assist with answering student questions, providing information, and managing routine tasks.

Benefits of AI Chatbots in Education

- **24/7 Availability:** Chatbots are available around the clock, offering support to students outside regular school hours.

- **Instant Responses:** Chatbots provide immediate answers to student queries, enhancing engagement and reducing wait times.

- **Consistency:** Chatbots deliver consistent responses, ensuring all students receive the same information.

- **Efficiency:** By handling routine tasks, chatbots free teachers to focus on more complex and personalized interactions.

Setting Up a Chatbot

Step 1: Define the Purpose

Determine the primary purpose of the chatbot. Will it answer student questions, provide homework help, or manage administrative tasks? For example, a chatbot could be designed to help students with math homework by answering common questions and providing step-by-step solutions.

Step 2: Choose a Platform

Select a platform to create and deploy your chatbot. Some popular options include:

- **Mitsuku:** An award-winning chatbot that can be customized for educational purposes.

- **Chatfuel:** A user-friendly platform that allows you to build chatbots for Facebook Messenger.

- **Dialogflow:** A Google-owned platform that supports various messaging platforms and offers advanced NLP capabilities.

Step 3: Develop the Chatbot

Develop the chatbot by creating conversation flows and responses. For example:

- **Greeting:** "Hello! I'm your AI assistant. How can I help you today?"

- **Homework Help:** "I can help you with your math homework. What problem are you working on?"

- **FAQs:** "Here are some common questions: 1. How do I log into the LMS? 2. Where can I find the syllabus?"

Step 4: Test and Iterate

Test the chatbot with a small group of students to identify any issues or areas for improvement. Gather feedback and make necessary adjustments. For example, if students find the chatbot's explanations too brief, you can expand the responses to provide more detailed information.

Step 5: Deploy the Chatbot

Deploy the chatbot across your chosen platforms, such as the school's website, LMS, or messaging apps. Ensure that students know how to access and use the chatbot.

Practical Applications of AI Chatbots

Homework Assistance

AI chatbots can provide instant homework help, guiding students through problems and offering explanations. For example, a math chatbot can help students solve algebraic equations by breaking down each step and providing hints.

Administrative Support

Chatbots can handle administrative tasks like answering questions about school policies, schedules, and events. They can also provide information on upcoming exams, deadlines, and school activities.

Language Learning

Language learning chatbots can engage students in conversation practice, helping them improve their language skills. For example, a Spanish language chatbot can converse with students and correct grammar and vocabulary usage.

Mental Health Support

AI chatbots can offer mental health support by providing resources and information on stress management, anxiety, and other mental health topics and resources. For example, a chatbot can suggest relaxation techniques or connect students with school counselors and helplines.

Setting Realistic Expectations

While AI chatbots are powerful tools, it's crucial to set realistic expectations for their capabilities. For instance, while they can provide general information and basic problem-solving, they may not effectively handle complex queries or emotional support.

Continuous Improvement

Continuously improve the chatbot based on student feedback and evolving needs. Regularly update the chatbot's knowledge base and conversation flows to remain relevant and helpful.

4.2 Using AI for Grading and Feedback: Pros, Cons, and Tips

AI-powered grading tools can significantly reduce the time and effort required for grading assignments, providing immediate feedback to students. This section examines the potential benefits and challenges of using AI for grading and offers practical tips for effective implementation.

Benefits of AI in Grading

Efficiency

AI grading tools can process and evaluate assignments quickly, reducing teachers' time on grading. For example, an AI-powered essay grading tool can analyze and score hundreds of essays in a fraction of the time it would take a human grader.

Consistency

AI grading tools provide consistent evaluations, eliminating potential biases that may arise from human grading. For example, an AI tool will apply the same criteria to all assignments, ensuring fair and objective assessments.

Immediate Feedback

AI grading tools offer instant feedback, allowing students to learn from their mistakes and improve. Grammar-checking tools like Grammarly provide real-time corrections and suggestions, helping students enhance their writing skills.

Potential Drawbacks

Limited Understanding

AI grading tools may struggle with nuanced or creative responses that require more profound understanding. For example, an AI tool might misinterpret the tone or style of a creative writing piece, leading to inaccurate evaluations.

Dependence on Training Data

The accuracy of AI grading tools depends on the training data's quality, integrity, and diversity. The tool's evaluations may be flawed if the training data is limited or skewed. For example, an AI tool trained on essays from native English speakers may not accurately evaluate essays from non-native speakers.

Lack of Personal Touch

AI grading tools lack the personal touch that human graders provide, such as encouraging comments and personalized feedback. For example, a human grader might offer suggestions for improvement and encouragement, which an AI tool cannot replicate.

Tips for Effective Implementation

Complement, Not Replace

Use AI grading tools to complement, not replace, human grading. Combine AI evaluations with human reviews to ensure accuracy and provide personalized feedback. For example, use an AI tool to grade grammar and syntax and review the content and creativity yourself.

Choose the Right Tool

Select AI grading tools that align with your grading criteria and subject matter. For example, specialized tools like Gradescope for math and science assignments and Grammarly for writing assignments can be used.

Set Clear Criteria

Define clear grading criteria for the AI tool to follow. For example, specify the importance of grammar, coherence, and content accuracy for an essay assignment. This helps the AI tool provide more accurate and relevant evaluations.

Monitor and Adjust

Continuously monitor the performance of AI grading tools and make necessary adjustments. Gather feedback from students and analyze the tool's evaluations to identify areas for improvement. For example, if the AI tool consistently misinterprets certain types of responses, adjust the training data or grading criteria.

Case Studies

Case Study 1: Automated Essay Grading with Gradescope

Context: A university implemented Gradescope to automate the grading of essays in an extensive introductory course.

Implementation: Students submitted essays through Gradescope, and the AI tool evaluated grammar, structure, and content based on predefined criteria.

Outcome: Teachers saved significant time on grading, allowing them to focus on sharing personalized input, support, and feedback with students. Students received instant feedback on their writing, helping them improve their skills throughout the course.

Case Study 2: Real-Time Feedback with Grammarly

Context: A high school English teacher used Grammarly to provide real-time feedback on student writing assignments.

Implementation: Students wrote their essays using Grammarly, which provided instant corrections and suggestions for improvement.

Outcome: Students improved their writing skills through immediate, personalized feedback. The teacher used Grammarly's evaluations to identify and address common issues in class.

4.3 AI-Driven Classroom Analytics for Enhanced Teaching Insights

AI-driven classroom analytics can provide valuable insights into student performance, engagement, and learning patterns. This section explores how AI analytics tools can enhance teaching and offers practical examples and case studies.

Understanding Classroom Analytics

What is Classroom Analytics?

Classroom analytics involve collecting and analyzing student performance, behavior, and engagement data. AI-driven analytics apply machine learning algorithms to identify patterns and trends, providing actionable insights for educators.

Benefits of AI-Driven Analytics

- **Personalized Learning:** AI analytics tools can identify individual student needs and preferences, enabling personalized instruction.

- **Early Intervention:** AI tools can detect early signs of academic struggles or disengagement, allowing for timely intervention.

- **Informed Decision-Making:** Analytics provide educators with data-driven, informed decisions about instructional strategies and resource allocation.

- **Continuous Improvement:** Regular analysis of classroom data helps educators refine their teaching methods and improve student outcomes.

Tools and Technologies

Learning Management Systems (LMS

Modern LMS platforms like Canvas and Blackboard incorporate AI-driven analytics to track student performance and engagement. For example:

- **Canvas Analytics:** Provides insights into student activity, assignment submissions, and grades, helping educators identify at-risk students and adjust instruction accordingly.

- **Blackboard Predict:** Uses predictive analytics to forecast student outcomes and suggest interventions to improve retention and success.

Adaptive Learning Platforms

Adaptive learning platforms like Knewton and DreamBox use AI analytics to tailor content to individual student needs. For example:

- **Knewton:** Analyzes student performance data to provide personalized coursework and identify areas for improvement.

- **DreamBox:** Tracks student progress in real-time and adjusts the difficulty of math exercises based on individual performance.

Classroom Management Tools

Classroom management tools like ClassDojo and Edmodo use AI analytics to monitor student behavior and engagement. For example:

- **ClassDojo:** Tracks student behavior and engagement, providing insights into class dynamics and individual student needs.

- **Edmodo:** Analyzes student interactions and participation, helping educators identify trends and tailor instruction.

Practical Applications

Identifying At-Risk Students

AI-driven analytics tools can identify at-risk students based on performance data, attendance records, and engagement metrics. For example:

- **Predictive Analytics:** Use tools like Blackboard Predict to forecast student outcomes and identify those at risk of failing or dropping out. Intervene early with targeted support and resources.

- **Behavior Tracking:** Use ClassDojo to monitor and identify patterns of student behavior that may indicate disengagement or academic struggles. Implement behavior interventions and support strategies.

Personalizing Instruction

AI analytics tools can help personalize instruction by identifying individual learning needs and preferences. For example:

- **Adaptive Learning:** Use Knewton to analyze student performance and provide personalized coursework. Tailor teaching to address specific challenge areas and build on strengths.

- **Real-Time Feedback:** Use DreamBox to track student progress and adjust the difficulty of math exercises in real-time. Provide immediate feedback and support to help students master concepts.

Improving Student Engagement

AI analytics tools can enhance student engagement by identifying factors influencing participation and motivation. For example:

- **Engagement Metrics:** Use Canvas Analytics to track student activity and participation in online discussions. Identify trends and implement strategies to boost engagement, such as interactive activities and collaborative projects.

- **Behavior Insights:** Use ClassDojo to monitor student behavior and engagement in class. Identify factors contributing to positive behavior and motivation and incorporate these elements into your teaching.

Case Studies

Case Study 1: Personalized Learning with Knewton

Context: A middle school implemented Knewton to personalize math instruction for students with diverse learning needs.

Implementation: Knewton analyzed student performance data and provided personalized coursework tailored to individual strengths and weaknesses.

Outcome: Students showed significant improvement in math proficiency, with many achieving above-grade-level performance. Teachers reported that Knewton helped them effectively identify and address individual learning needs.

Case Study 2: Early Intervention with Blackboard Predict

Context: A university used Blackboard Predict to identify at-risk students in an extensive introductory course.

Implementation: Blackboard Predict analyzed student performance, attendance, and engagement data to forecast outcomes and identify at-risk students.

Outcome: The university implemented targeted interventions for at-risk students, such as tutoring and academic support. As a result, student retention and success rates improved significantly.

4.4 Enhancing Collaboration and Communication with AI

AI tools can enhance collaboration and strengthen communication among students, teachers, and parents. This section explores how AI-driven platforms and applications can facilitate effective communication and collaborative learning.

AI for Student Collaboration

Collaborative Learning Platforms

AI-driven collaborative learning platforms like Microsoft Teams and Google Classroom support student collaboration through shared workspaces, real-time communication, and project management tools. For example:

- **Microsoft Teams:** Provides a collaborative workspace where students can work on group projects, share files, and communicate in real-time. AI features like translation and transcription facilitate communication among diverse student groups.

- **Google Classroom:** Offers tools for creating and managing group assignments, facilitating discussions, and sharing resources. AI features, such as automated grading and personalized feedback, enhance collaboration.

Virtual Study Groups

AI-driven virtual study groups connect students with peers for collaborative learning and support. For example:

- **Brainly:** An AI-powered platform where students can ask questions and receive answers from peers. AI features help identify the best answers and provide additional resources.

- **Quizlet:** Uses AI to create personalized study sets and connect students with study partners for collaborative learning.

Collaborative Projects

AI tools can enhance collaborative projects by providing real-time feedback and support. For example:

- **CoWriter:** An AI-driven writing assistant that helps students collaborate on writing projects. Provides real-time feedback on grammar, style, and coherence, facilitating effective collaboration.

- **Mentimeter:** An interactive presentation tool that uses AI to gather real-time student feedback and insights during collaborative projects.

AI for Teacher-Student Communication

Virtual Office Hours

AI-driven platforms can facilitate virtual office hours, allowing teachers to provide support and answer questions outside of regular class time. For example:

- **Zoom:** Offers AI features, such as automated scheduling and transcription, to streamline virtual office hours and enhance communication.

- **Google Meet:** Provides tools for scheduling and conducting virtual meetings, with AI features for real-time translation and transcription.

AI-Powered Feedback

AI tools can provide personalized feedback on assignments and assessments, enhancing communication between teachers and students. For example:

- **Turnitin:** An AI-powered plagiarism detection and feedback tool that provides detailed analysis and suggestions for improvement.

- **Perusall:** An AI-driven platform that facilitates collaborative annotation and discussion of texts, providing real-time feedback and insights.

AI for Parent-Teacher Communication

Automated Updates

AI-driven platforms can automate updates and communications, informing parents about student progress and school activities. For example:

- **Remind** An AI-powered communication tool that sends automated updates and reminders to parents about assignments, events, and student progress.

- **Bloomz:** Provides tools for scheduling and managing parent-teacher conferences, with AI features for automated reminders and communication.

Data-Driven Insights

AI analytics tools can give parents data-driven insights about their child's performance and progress. For example:

- **Schoolytics:** An AI-powered platform aggregating and analyzing student performance data, providing parents with detailed reports and insights.

- **ClassTag:** Uses AI to track and analyze student engagement and performance, offering personalized insights and recommendations for parents.

Case Studies

Case Study 1: Collaborative Learning with Microsoft Teams

Context: A high school implemented Microsoft Teams to support collaborative learning in a remote learning environment.

Implementation: Students used Microsoft Teams to work on group projects, share files, and communicate in real-time. AI features like translation and transcription facilitated communication among diverse student groups.

Outcome: Students reported improved collaboration and communication, leading to higher-quality group projects and increased engagement. Teachers found that Microsoft Teams enhanced their ability to support and monitor student collaboration.

Case Study 2: Virtual Office Hours with Zoom

Context: A university used Zoom to conduct virtual office hours and provide additional student support.

Implementation: Professors scheduled virtual office hours using Zoom, with AI features for automated scheduling and transcription.

Outcome: Students appreciated the flexibility and accessibility of virtual office hours, leading to increased participation and support. Professors reported that AI features streamlined the process and enhanced communication.

Conclusion

Integrating AI into classroom management and efficiency offers numerous benefits, from acting as virtual teaching assistants and providing real-time feedback to enhancing collaboration and communication. Educators can create more effective, personalized, and engaging learning environments by thoughtfully implementing AI tools. The examples and case studies in this chapter illustrate the transformative potential of AI in education and offer practical strategies for successful integration. As educators continue to explore and adopt AI technologies, they can unlock new possibilities for teaching and learning, ultimately benefiting students, teachers, and the broader educational community.

Chapter 5: Subject-Specific AI Applications

5.1 Enhancing STEM Education with AI Tools

AI tools that make complex concepts more accessible and engaging can greatly benefit STEM (science, technology, engineering, and mathematics) education. This section explores various AI applications in STEM education, offering practical examples and success stories. By integrating these tools, educators can empower themselves with the confidence and knowledge needed to deliver more engaging and personalized learning experiences.

Interactive Simulations

AI-driven interactive simulations provide students hands-on experiences that make abstract STEM concepts tangible and understandable.

Example: PhET Interactive Simulations

PhET offers a suite of interactive simulations for subjects like physics, chemistry, and biology. These simulations, powered by AI algorithms, allow students to experiment with different variables and see real-time results. For example, a simulation of electric circuits enables students to build virtual circuits, adjust components, and observe the effects on current and voltage. The AI technology behind these simulations enhances the learning experience by providing personalized feedback and adapting the simulation to the student's progress.

Case Study: Using PhET in High School Physics

Context: A high school physics teacher used PhET simulations to teach electric circuits.

Implementation: The teacher integrated PhET simulations into the lesson plan, allowing students to virtually explore circuit components and configurations.

Outcome: Students reported a deeper understanding of electric circuits and increased engagement. The interactive nature of the simulations, powered by AI, helped them grasp complex concepts that were challenging to visualize through traditional methods. This enhanced engagement can lead to better learning outcomes.

Personalized Learning Paths

AI-powered adaptive learning platforms can tailor STEM instruction to individual student needs, providing personalized learning paths that address their strengths and weaknesses. However, it's important to consider the ethical implications of using AI in education, such as data privacy and student autonomy. This section explores various AI applications in STEM education, offering practical examples and success stories, while also discussing these ethical considerations.

Example: DreamBox Learning for Math

DreamBox Learning uses AI to adapt math lessons to each student's proficiency level. The platform continuously assesses student performance and adjusts the difficulty of exercises accordingly. The AI's feedback mechanism is designed to be accurate and comprehensive, providing students with a clear understanding of their strengths and areas for improvement.

Case Study: DreamBox Learning in Middle School Math

Context: A middle school implemented DreamBox Learning to personalize math instruction for students with diverse learning needs. The AI platform's adaptive learning features have been particularly beneficial for students with special needs, allowing them to learn at their own pace and providing additional support when needed.

Implementation: Teachers assigned math exercises on DreamBox, and the AI platform adjusted the difficulty based on individual performance.

Outcome: Students showed significant improvement in math proficiency, with many advancing beyond grade-level expectations. Teachers reported that DreamBox, with its AI-powered adaptive learning, helped them effectively identify and address individual learning needs, fostering each student's unique growth.

Automated Problem-Solving

AI tools can assist in teaching computational thinking and problem-solving skills through coding and robotics.

Example: Code.org and AI-Driven Coding Platforms

Code.org offers AI-driven coding platforms that guide students through coding exercises and projects. The AI provides real-time feedback and suggestions, helping students debug their code and improve their programming skills.

Case Study: Code.org in Elementary Schools

Context: An elementary school integrated Code.org into its computer science curriculum to teach basic coding skills.

Implementation: Students used Code.org to complete coding exercises and projects, with the AI providing real-time feedback and support.

Outcome: Students developed strong foundational coding skills and demonstrated increased interest in computer science. The AI's real-time feedback helped them learn from their mistakes and improve their problem-solving abilities.

Data Analysis Projects

AI tools can enable students to analyze sophisticated data, fostering critical thinking in science and math.

Example: IBM Watson Studio for Education

IBM Watson Studio offers data analysis and visualization tools, allowing students to work on real-world data projects. The AI platform provides insights and recommendations, helping students analyze data and draw meaningful conclusions.

Case Study: Data Analysis with IBM Watson Studio

Context: A high school science class used IBM Watson Studio to analyze environmental data.

Implementation: Students collected data on local air quality and used IBM Watson Studio to analyze trends and correlations. The AI provided insights and visualizations to support their analysis.

Outcome: Students gained hands-on experience with data analysis and developed critical thinking skills. The project fostered a deeper understanding of environmental science and the importance of data in scientific research.

5.2 AI in Language Arts: From Automated Essay Scoring to Creative Writing Prompts

Language arts education can benefit from AI tools that enhance reading comprehension, writing skills, and literary analysis. This section explores various AI applications in language arts education, providing practical examples and success stories.

Enhanced Reading Comprehension

AI applications can adapt reading materials to individual student levels and provide real-time comprehension assistance.

Example: Lexia Learning

Lexia Learning offers AI-driven reading programs that assess student reading levels and provide personalized reading practice. The AI adjusts the difficulty of texts and offers real-time feedback on comprehension.

Case Study: Lexia Learning in Elementary Schools

Context: An elementary school implemented Lexia Learning to support struggling readers.

Implementation: Students used Lexia Learning to complete personalized reading exercises, with the AI adjusting the difficulty and providing feedback.

Outcome: Students showed significant improvement in reading fluency and comprehension. Teachers found that Lexia Learning helped them provide targeted support to struggling readers and track their progress.

Automated Writing Feedback

AI tools can offer immediate, personalized feedback on student writing, from grammar to style.

Example: Grammarly

Grammarly provides real-time grammar, style, and coherence feedback, helping students improve their writing skills.

Case Study: Grammarly in High School English Classes

Context: A high school English teacher used Grammarly to provide real-time feedback on student essays.

Implementation: Students wrote their essays using Grammarly, which provided instant corrections and suggestions for improvement.

Outcome: Students improved their writing skills through immediate, personalized feedback. The teacher used Grammarly's evaluations to identify and address common issues in class.

Creative Writing with AI

AI tools can generate creative writing prompts and assist students in overcoming writer's block. It's important to note that these AI tools are not meant to replace creativity, but rather to stimulate it. They are designed to provide a starting point for creative thinking and to support students in their creative process.

Example: Plot Generator

Plot Generator uses AI to create unique writing prompts and story ideas, helping students get started on their creative writing projects.

Case Study: Plot Generator in Middle School Creative Writing

Context: A middle school implemented a Plot Generator to support creative writing assignments.

Implementation: Students used Plot Generator to generate writing prompts and story ideas. The AI provided suggestions for characters, settings, and plot twists.

Outcome: Students produced more creative and engaging stories, with many reporting that the AI prompts helped them overcome writer's block. Teachers found that Plot Generator encouraged students to explore new ideas and improve their creative writing skills.

Analyzing Literature with AI

AI tools can help students analyze themes, motifs, and symbols in literature, making complex analyses more accessible.

Example: LitCharts

LitCharts uses AI to provide detailed analyses of literary works, including themes, motifs, and symbols. The AI technology behind LitCharts has been developed and tested by a team of experts in literature and AI, ensuring its reliability and validity. The platform is continuously updated based on user feedback, the latest trends, and research.

Case Study: LitCharts in High School Literature Classes

Context: A high school literature teacher used LitCharts to support the analysis of classic novels.

Implementation: Students used LitCharts to explore the themes, motifs, and symbols in novels like "To Kill a Mockingbird" and "1984." The AI provided detailed explanations and examples.

Outcome: Students better understood the literary works and produced more insightful analyses. Teachers found that LitCharts helped students engage with the material and develop critical thinking skills.

5.3 Bringing History to Life with AI Simulations

AI-powered simulations that immerse students in historical periods and events can transform history education. This section explores various AI applications in history education, offering practical examples and success stories.

Historical Simulations

AI-powered simulations can immerse students in historical periods, enhancing understanding and engagement.

Example: Mission US

Mission US offers AI-driven interactive simulations that allow students to experience critical moments in American history.

Case Study: Mission US in Middle School History Classes

Context: A middle school implemented Mission US to teach American history.

Implementation: Students participated in interactive simulations that placed them in historical scenarios, such as the American Revolution and the Civil Rights Movement.

Outcome: Students reported increased engagement and a deeper understanding of historical events. Teachers found that the simulations helped students empathize with historical figures and grasp the complexities of historical contexts.

Analyzing Historical Data

AI tools can help students analyze and interpret historical data and trends.

Example: Google Arts & Culture

Google Arts & Culture provides AI-driven tools for exploring historical artifacts, documents, and timelines.

Case Study: Google Arts & Culture in High School History Classes

Context: A high school history teacher used Google Arts & Culture to analyze historical data and artifacts.

Implementation: Students used the platform to explore primary sources, such as letters, photographs, and maps. The AI provided context and analysis, helping students interpret the data.

Outcome: Students developed critical thinking skills and a deeper understanding of historical research. Teachers found that Google Arts & Culture made historical data more accessible and engaging for students.

Virtual Field Trips

AI-driven virtual reality (VR) experiences can transport students to historical sites and events.

Example: Google Expeditions

Google Expeditions offers VR field trips that allow students to explore historical sites and events.

Case Study: Google Expeditions in Middle School History Classes

Context: A middle school used Google Expeditions to take virtual field trips to historical sites.

Implementation: Students explored ancient Rome, the Great Wall of China, and the Normandy beaches. The AI provided guided tours and interactive elements.

Outcome: Students reported increased interest and engagement in history. Teachers found that the VR experiences helped students visualize and connect with historical events more effectively.

Interactive Timelines

AI can create interactive timelines where students explore historical events and their interconnectedness.

Example: Tiki-Toki

Tiki-Toki uses AI to create interactive multimedia timelines illustrating historical events and their relationships.

Case Study: Tiki-Toki in High School History Classes

Context: A high school history teacher used Tiki-Toki to create interactive timelines for studying World War II.

Implementation: Students explored timelines with multimedia elements, such as videos, photos, and primary source documents. The AI provided context and connections between events.

Outcome: Students demonstrated a deeper understanding of the chronology and interconnectedness of historical events. Teachers found that Tiki-Toki helped students engage with the material and develop critical thinking skills.

5.4 AI for Art Education: Exploring Digital Creativity

AI tools that enhance digital creativity and provide interactive, personalized learning experiences can benefit art education. This section explores AI applications in art education, offering practical examples and success stories.

AI in Digital Art Creation

AI tools can assist students in creating digital art, offering new possibilities for creativity and expression.

Example: DeepArt

DeepArt uses AI to transform photos into artworks in the style of renowned artists, such as Picasso and Van Gogh.

Case Study: DeepArt in High School Art Classes

Context: A high school art teacher used DeepArt to introduce students to digital art creation.

Implementation: Students used DeepArt to transform their photos into artworks in different artistic styles. The AI provided options for customization and experimentation.

Outcome: Students explored new forms of digital creativity and produced unique artworks. Teachers found that DeepArt encouraged students to experiment with different styles and techniques.

Interactive Art History Lessons

AI tools can personalize art history lessons, making them more engaging and accessible.

Example: SmartHistory

SmartHistory uses AI to provide interactive, multimedia art history lessons.

Case Study: SmartHistory in Middle School Art Classes

Context: A middle school art teacher used SmartHistory to teach art history.

Implementation: Students used SmartHistory to explore interactive lessons on various art movements and artists. The AI provided multimedia content, such as videos, images, and quizzes.

Outcome: Students demonstrated increased interest and engagement in art history. Teachers found that SmartHistory made art history more accessible and interactive for students.

Critiquing and Analyzing Art with AI

AI tools can help students critique and analyze artworks, fostering a deeper understanding.

Example: Artivive

Artivive uses AI to overlay digital information on physical artworks, enhancing analysis and critique.

Case Study: Artivive in High School Art Classes

Context: A high school art teacher used Artivive to support art critique and analysis.

Implementation: Students used Artivive to overlay digital information on artworks, providing context, analysis, and critique. The AI offered insights into artistic techniques, historical context, and symbolism.

Outcome: Students developed critical thinking skills and a deeper understanding of art analysis. Teachers found that Artivive enhanced the critique process and encouraged deeper engagement with artworks.

Augmented Reality in Art

Augmented reality (AR) apps can create interactive art projects that blend physical and digital media.

Example: QuiverVision

QuiverVision offers AR apps that bring coloring pages to life, allowing students to create interactive art projects.

Case Study: QuiverVision in Elementary School Art Classes

Context: An elementary school used QuiverVision to introduce students to augmented reality art projects.

Implementation: Students colored pages provided by QuiverVision and used the AR app to bring their creations to life. The AI added interactive elements, such as animation and sound.

Outcome: Students demonstrated increased interest and engagement in art projects. Teachers found that QuiverVision encouraged creativity and experimentation with new media.

Conclusion

AI offers transformative potential for subject-specific applications in education. AI tools can enhance STEM education, language arts, history, and art to create more engaging, personalized, and compelling learning experiences. The examples and case studies in this chapter illustrate the diverse ways AI can be integrated into various subjects, offering practical strategies for successful implementation. As educators continue exploring and adopting AI technologies, they can unlock new possibilities for teaching and learning, benefiting students and enriching the educational experience.

8

Chapter 6: Engaging Students with AI

A I, or Artificial Intelligence, is a beacon of innovation that has the transformative potential to enhance student engagement in the classroom significantly. By creating interactive, immersive, and personalized learning experiences, AI opens up a world of possibilities for student learning. This chapter explores various AI applications that can revolutionize learning, providing detailed examples and case studies to ignite your optimism about the future of education.

6.1 Gamification of Learning through AI

Gamification applies game design qualities to non-game contexts to motivate and engage learners. With its ability to personalize the experience and provide real-time feedback, AI can ignite excitement and curiosity in students, making learning a thrilling adventure.

Benefits of Gamification

- **Increased Motivation:** Game elements such as points, badges, and leaderboards can motivate students to engage more actively in their learning.

- **Enhanced Engagement:** Interactive and competitive elements make learning more engaging and enjoyable. AI-driven gamification provides instant feedback, helping students track their real-time progress and identify areas for improvement. This sense of achievement and progress can be very inspiring, guiding students' learning process in real-time and fostering a sense of accomplishment.

- **Personalization:** AI can tailor game-based learning experiences to individual student needs and preferences.

Examples of AI-Driven Gamification

Example 1: Kahoot!

Kahoot! is a popular game-based learning platform that uses AI to create interactive quizzes and games. The AI in Kahoot! is designed to personalize the learning experience by adapting the difficulty of the questions based on the student's performance. Teachers can create quizzes on any subject and invite students to complete them in real-time.

Implementation:

- **Quiz Creation:** Teachers create quizzes on the Kahoot! Platform, incorporating multimedia elements such as images and videos.

- **Real-Time Play:** Students join the game using a unique code and compete to answer questions correctly and quickly.

- **Feedback and Analytics:** The AI provides real-time feedback on student performance and offers analytics to help teachers understand student progress.

Case Study: Kahoot! in High School Biology

Context: In a high school biology class, the teacher used Kahoot!, a popular game-based learning platform that uses AI, to review material before exams.

Implementation: The teacher created interactive quizzes covering fundamental concepts in biology. Students competed in real-time, answering questions on cell structure, genetics, and evolution.

Outcome: Students reported increased engagement and enjoyment. The competitive nature of Kahoot! Motivated them to review the material thoroughly. The teacher used the analytics to identify areas where students needed additional support.

Example 2: Classcraft

Classcraft gamifies the classroom experience into a role-playing game (RPG). Students create characters and earn points for positive behaviors and academic achievements.

Implementation:

- **Character Creation:** Students create characters and choose roles such as warrior, mage, or healer.

- **Earning Points:** Students earn points for completing assignments, participating in class, and demonstrating positive behavior.

- **Quests and Challenges:** The AI in Classcraft generates quests and challenges that align with the curriculum and provide opportunities for collaborative learning. These quests and challenges are designed to engage students and improve their comprehension of the subject matter.

Case Study: Classcraft in Middle School English

Context: A middle school English teacher used Classcraft to motivate students and improve classroom behavior.

Implementation: Students created characters and earned points for completing reading assignments, participating in discussions, and submitting essays on time. The AI-generated quests related to the literature they were studying.

Outcome: Students were more motivated to complete their work and participate in class. The game elements fostered a sense of community and collaboration. The teacher observed improvements in both academic performance and classroom behavior.

6.2 Virtual Reality (VR) and Augmented Reality (AR) in the Classroom

VR and AR technologies can create immersive learning experiences that enhance student engagement and understanding.

Benefits of VR and AR

- **Immersive Learning:** VR and AR create immersive environments that make learning more engaging and memorable.

- **Enhanced Understanding:** These technologies can make abstract or complex concepts more concrete and accessible.

- **Interactivity:** VR and AR allow interactive learning experiences where learners can interact with and manipulate virtual objects.

- **Personalization:** AI can personalize VR and AR experiences by adapting the content and difficulty level according to student performance and preferences, making the learning experience more effective and engaging.

Examples of VR and AR in Education

Example 1: Google Expeditions

Google Expeditions is a VR and AR platform enabling educators to take students on virtual field trips.

Implementation:

- **Virtual Field Trips:** Teachers can choose from various virtual field trips, from exploring ancient civilizations touring modern science labs.

- **Guided Tours:** The platform provides detailed information and interactive elements.

- **Student Exploration:** Students can explore virtual environments using VR headsets or AR-enabled devices.

Case Study: Google Expeditions in Elementary School Geography

Context: An elementary school geography teacher used Google Expeditions to teach about different biomes.

Implementation: Students took virtual field trips to various biomes, such as the Amazon rainforest, the Sahara desert, and the Arctic tundra. They explored the unique characteristics of each biome and interacted with virtual flora and fauna.

Outcome: Students demonstrated a deeper understanding of biomes and were highly engaged in the lessons. The immersive experience made the material more memorable and fostered a greater interest in geography.

Example 2: zSpace

zSpace is an AR platform that provides interactive, 3D learning experiences in science, math, and engineering subjects.

Implementation:

- **Interactive Lessons:** Teachers use zSpace to deliver interactive lessons that allow students to manipulate 3D models and conduct virtual experiments.

- **Real-Time Feedback**: The AI provides real-time feedback and guidance, helping students understand complex concepts.

- **Collaborative Learning:** Students can collaborate on projects and share their findings using the AR platform.

Case Study: zSpace in High School Physics

Context: A high school physics teacher used zSpace to teach concepts such as force and motion.

Implementation: Students conducted virtual experiments using 3D models of pulleys, levers, and inclined planes. They manipulated the models to observe the effects of different forces and recorded their observations.

Outcome: Students showed an improved understanding of physics concepts and increased engagement in the subject. The interactive nature of zSpace made the lessons more enjoyable and accessible.

6.3 AI-Powered Tools for Interactive Science Experiments

AI-powered tools, such as virtual labs, can facilitate interactive science experiments. These tools use AI to provide real-time feedback, assist with data collection and analysis, and ensure safety and accessibility, making it easier for students to engage with scientific concepts and develop critical thinking skills.

Benefits of AI-Powered Science Experiments

- **Hands-On Learning:** AI tools enable hands-on learning experiences that enhance understanding and retention.

- **Real-Time Feedback:** AI provides immediate feedback on students' experiments, helping them learn from their mistakes and refine their scientific skills in real-time.

- **Safety and Accessibility:** Virtual experiments eliminate safety risks and make complex experiments more accessible.

- **Data Analysis:** AI can assist with data collection and analysis in science experiments, providing insights that support scientific inquiry and helping students understand and interpret their experimental results.

Examples of AI-Powered Science Experiments

Example 1: Labster

Labster offers a virtual lab platform that allows students to conduct interactive science experiments.

Implementation:

- **Virtual Labs:** Students conduct virtual biology, chemistry, and physics experiments. The AI guides them through the process and provides real-time feedback.

- **Simulations:** The platform offers realistic simulations of lab equipment and procedures, making it possible to conduct complex experiments safely.

- **Data Analysis:** AI tools assist with data collection and analysis, helping students conclude from their experiments.

Case Study: Labster in High School Biology

Context: A high school biology teacher used Labster to conduct virtual lab experiments on cellular respiration.

Implementation: Students measured cellular respiration in yeast cells using a virtual experiment. They manipulated variables such as temperature and glucose concentration and recorded their observations.

Outcome: Students better understood cellular respiration and the scientific method. The virtual lab experience made the experiment more accessible and engaging without the safety concerns of a physical lab.

Example 2: PhET Interactive Simulations

PhET Interactive Simulations offers AI-driven simulations for a wide range of science topics.

Implementation:

- **Interactive Simulations:** Students interact with virtual simulations to explore scientific concepts hands-on.

- **Guided Experiments:** The AI guides students through experiments, providing instructions and feedback.

- **Visualization:** The simulations offer visual representations of abstract concepts, making them easier to understand.

Case Study: PhET in Middle School Chemistry

Context: A middle school chemistry teacher used PhET simulations to teach about chemical reactions.

Implementation: Students used PhET to simulate chemical reactions, mix virtual chemicals, and observe the results. The AI provided real-time feedback and explanations of the responses.

Outcome: Students demonstrated an improved understanding of chemical reactions and increased engagement in the subject. The interactive simulations made the abstract concepts more tangible and accessible.

6.4 Building AI Literacy: Projects and Activities for Students

Developing AI literacy is essential for preparing students for a future where AI plays a significant role. This section explores projects and activities that help students understand AI and its applications.

Understanding AI

Teaching AI Basics

Introduce students to the basics of AI, including deep learning, gamification, and neural networks.

Example: AI for Everyone by Coursera

AI for Everyone is an online course that provides a non-technical introduction to AI.
  Implementation:

- **Course Enrollment:** Students enroll in the course and complete modules on AI concepts and applications.

- **Interactive Lessons:** The course includes interactive lessons and quizzes to reinforce learning.

- **Project Work:** Students apply their knowledge by working on projects related to AI applications.

Case Study: AI for Everyone in High School Computer Science

Context: A high school computer science teacher used AI for Everyone to introduce students to AI concepts.

Implementation: Students completed the online course and worked on projects to apply their knowledge. For example, they developed simple AI models using available datasets.

Outcome: Students gained a foundational understanding of AI and its applications. The projects helped them see its real-world relevance and sparked their interest in further exploration.

Ethical Considerations of AI

Exploring AI Ethics

Discuss the implications of AI on ethics, including issues such as AI data biases, risks to privacy, and the impact on jobs and society.

Example: AI and Ethics Workshops

Organize workshops to explore the ethical considerations of AI.

Implementation:

- **Guest Speakers:** Invite experts in AI ethics to speak to students about the ethical challenges and considerations.

- **Case Studies:** Present case studies of AI applications with ethical implications, such as facial recognition and autonomous vehicles.

- **Debates and Discussions:** Facilitate debates and discussions on ethical issues, encouraging students to think critically and form their opinions.

Case Study: AI Ethics Workshop in High School Social Studies

Context: A high school social studies teacher organized an AI ethics workshop.

Implementation: The workshop included guest speakers, case studies, and student debates on AI-related ethical issues. For example, students debated using AI in surveillance and its privacy implications.

Outcome: Students developed a deeper understanding of AI's ethical considerations. The workshop encouraged critical thinking and moral reasoning, equipping students with the skills necessary to comprehend the complex landscape of AI in society.

Hands-On AI Projects

Creating AI Projects

Engage students in hands-on projects that involve developing and applying AI models.

Example: AI Robotics Projects

Organize robotics projects that involve programming and training AI models.

Implementation:

- **Robotics Kits:** Provide students with robotics kits with sensors and programmable components.

- **AI Programming:** Teach students to program AI models using tools like TensorFlow or Python.

- **Project-Based Learning:** Students work on projects to develop AI-driven robots, such as autonomous vehicles or robotic assistants.

Case Study: AI Robotics Project in Middle School STEM Club

Context: A middle school STEM club engaged students in an AI robotics project.

Implementation: Students used robotics kits and programming tools to develop AI-driven robots. For example, they built and programmed robots to navigate mazes autonomously.

Outcome: Students gained practical experience with AI and robotics. The project fostered creativity, problem-solving skills, and a deeper understanding of AI applications.

Conclusion

AI offers transformative potential for enhancing student engagement in education. By leveraging gamification, VR and AR, interactive science experiments, and AI literacy projects, educators can create dynamic and immersive learning experiences. The examples and case studies in this chapter illustrate the diverse ways AI can be integrated into educational environments to engage students and enrich their academic journey. As educators continue to explore and adopt AI technologies, they can unlock new possibilities for teaching and learning, ultimately benefiting students and preparing them for a future where AI plays a significant role.

9

Chapter 7: Addressing Diverse Learning Needs with AI

A I technology, a powerful tool in the hands of educators and administrators, can revolutionize education by addressing diverse learning needs and providing personalized support to students with varying abilities and challenges. This chapter delves into the ways AI can empower you to support reading interventions, customize math education, assist English language learners, and provide tools for special education.

7.1 AI-Assisted Reading Interventions for Struggling Students

Struggling readers often need targeted interventions to improve their skills and confidence. AI can provide personalized reading support that adapts to each student's needs.

Personalized Reading Support

AI-driven platforms can assess individual reading levels and provide tailored exercises and feedback.

Example: Amira Learning

Amira Learning is an AI-powered reading assistant that improves students' reading comprehension and fluency. The platform uses speech recognition to listen to students read aloud and provides instant feedback.

Implementation:

- **Assessment:** Amira Learning assesses each student's reading level by having them read a passage aloud. The AI analyzes the reading speed, accuracy, and expression to determine the appropriate level.

- **Personalized Practice:** Based on the assessment, the platform assigns personalized reading exercises that target areas for improvement, such as decoding, sight word recognition, or reading comprehension.

- **Real-Time Feedback:** As students read, the AI provides real-time feedback on pronunciation and fluency, helping them correct mistakes and improve.

Case Study: Amira Learning in Elementary Schools

Context: An elementary school implemented Amira Learning to support struggling readers in grades 1-3.

Implementation: Students used Amira Learning, which is available for a monthly subscription of [$ 10], for 15 minutes daily. The AI provided personalized reading exercises and real-time feedback.

Outcome: The implementation of Amira Learning in the elementary school led to a remarkable improvement in reading fluency and comprehension among students. Teachers were able to identify specific reading challenges and provide targeted support, thanks to the insights provided by Amira Learning. This success story is a testament to the transformative power of AI in education.

Interactive Reading Apps

Interactive reading apps use AI to engage students in reading practice tailored to their needs.

Example: Lexia Core5 Reading

Lexia Core5 Reading is an AI-driven reading program offering personalized learning paths for pre-K through 5th-grade students.

Implementation:

- **Initial Assessment:** Students complete an initial assessment to determine their reading level and identify areas for growth.

Personalized Activities: The AI assigns personalized reading activities that focus on phonemic skills, fluency, vocabulary, reading comprehension, and more.

- **Progress Monitoring:** The platform continuously monitors student progress and adjusts the difficulty of activities accordingly.

Case Study: Lexia Core5 Reading in Middle Schools

Context: A middle school used Lexia Core5 Reading to support students reading below grade level.

Implementation: Students used the platform for 20 minutes daily during a designated reading intervention period. Teachers used the progress reports to tailor their instruction and provide additional support.

Outcome: Students demonstrated significant gains in reading skills, closing the gap between their reading level and grade-level expectations. Teachers found that Lexia Core5 Reading provided valuable insights into student progress and areas needing further attention.

Speech Recognition for Reading Fluency

AI-powered speech recognition tools can help students improve their reading fluency by providing immediate feedback on pronunciation and pacing.

Example: ReadWorks

ReadWorks offers an AI-driven reading comprehension platform with tools for improving reading fluency.

Implementation:

- **Reading Practice:** Students read passages aloud using the ReadWorks platform. The AI listens to their reading and provides pronunciation, speed, and expression feedback.

- **Comprehension Questions:** After reading, students answer comprehension questions that the AI adapts based on their responses, ensuring a personalized learning experience.

Case Study: ReadWorks in Elementary Schools

Context: An elementary school implemented ReadWorks to enhance fluency in reading and comprehension skills for students in grades 2 5.

Implementation: Students used ReadWorks during their daily reading block. The AI provided feedback on their reading and adjusted the comprehension questions to match their skill level.

Outcome: Students showed improvement in reading fluency and comprehension. Teachers reported that the real-time feedback from the AI helped students become more confident readers.

AI and Dyslexia

AI technologies can assist students with dyslexia by providing customized reading experiences and tools to overcome specific challenges.

Example: Bookshare

Bookshare is an AI-powered platform that offers accessible eBooks for students with dyslexia and other reading disabilities.

Implementation:

- **Customized Reading Options:** The platform provides eBooks with customizable text and audio options, allowing students to adjust font size, background color, and voice speed.

- **Reading Tools:** AI-driven tools such as text-to-speech, synchronized highlighting, and dictionary support help students with dyslexia understand and retain information. For instance, the text-to-speech feature can read the text aloud, while the synchronized highlighting feature can highlight the words as they are read, aiding in word recognition and comprehension.

Case Study: Bookshare in Special Education

Context: A special education program implemented Bookshare to support students with dyslexia.

Implementation: Students used Bookshare to access their textbooks and reading materials. The AI tools helped them read and comprehend the content more effectively.

Outcome: Students demonstrated increased reading comprehension and engagement. Teachers reported that Bookshare made reading more accessible and enjoyable for students with dyslexia.

7.2 Customizing Math Education with Adaptive Learning Platforms

Adaptive learning platforms use AI to tailor math instruction to individual student needs, providing personalized support and challenging students appropriately.

Adaptive Learning Technologies

Adaptive learning platforms analyze student performance and adjust the content in real-time to ensure personalized instruction.

Example: DreamBox Learning

DreamBox Learning is an adaptive math platform that provides personalized lessons based on student performance.

Implementation:

- **Initial Assessment:** Students complete an initial assessment to determine their math proficiency. It's important to note that all student data is encrypted and stored securely, ensuring the privacy and security of student information.

- **Personalized Lessons:** The AI in DreamBox Learning is not just a tool, but a partner in education. It assigns personalized lessons that adapt in real-time based on each student's unique performance, providing targeted practice and feedback. This adaptability ensures that every student receives the support they need to excel in math.

- **Progress Monitoring:** Teachers can monitor student progress through the platform's analytics dashboard, identifying areas where students need additional support.

Case Study: DreamBox Learning in Middle Schools

Context: A middle school implemented DreamBox Learning to support students with diverse math abilities.

Implementation: After a brief training session, students used DreamBox Learning during their math class and for homework. The AI provided personalized lessons and real-time feedback, making the platform easy to integrate into the existing curriculum.

Outcome: Students showed significant improvement in math proficiency, with many advancing beyond grade-level expectations. Teachers reported that DreamBox Learning helped them effectively identify and address individual learning needs.

Visual and Interactive Math Tools

AI tools can provide visual and interactive math experiences that cater to different learning styles.

Example: GeoGebra

GeoGebra is an AI-powered math platform offering interactive geometry, algebra, and calculus tools.

Implementation:

- **Interactive Lessons:** Teachers use GeoGebra to create interactive math lessons that allow students to manipulate variables and visualize concepts.

- **Real-Time Feedback:** The AI in GeoGebra provides real-time feedback and hints. For instance, if a student is trying to solve an algebraic equation and makes a mistake, the AI will immediately point out the error and provide a hint on how to correct it, enhancing the student's learning experience.

- **Collaborative Learning:** Students can collaborate on projects and share their findings using the platform.

Case Study: GeoGebra in High School Math

Context: A high school math teacher used GeoGebra to teach geometry and algebra.

Implementation: Students used GeoGebra to explore geometric shapes, solve algebraic equations, and visualize calculus concepts. The AI provided real-time feedback and hints.

Outcome: Students demonstrated a deeper understanding of math concepts and increased engagement in the subject. The interactive nature of GeoGebra made the lessons more enjoyable and accessible.

Real-Time Feedback and Adjustments

AI-powered platforms can provide real-time feedback and adjust instruction based on student performance.

Example: Knewton Alta

Knewton Alta is an AI platform focused on adaptive learning that offers personalized coursework in various subjects, including math.

Implementation:

- **Initial Assessment:** Students complete an initial assessment to determine their math proficiency.

- **Personalized Coursework:** The AI assigns personalized coursework that adapts in real-time based on student performance, providing targeted practice and feedback.

- **Progress Monitoring:** Teachers can monitor student progress through the platform's analytics dashboard, identifying areas where students need additional support.

Case Study: Knewton Alta in College Math Courses

Context: A college math department implemented Knewton Alta to support students in introductory math courses.

Implementation: Students used Knewton Alta for coursework and homework. The AI provided personalized lessons and real-time feedback.

Outcome: Students demonstrated significant improvement in math proficiency and course completion rates. Teachers reported that Knewton Alta helped them effectively identify and address individual learning needs.

Engaging Math Challenges

AI-driven platforms can offer challenging math problems and puzzles that adapt to student skill levels.

Example: Prodigy Math

Prodigy Math is an AI-powered platform offering personalized math challenges and games for grades 1-8 students.

Implementation:

- **Personalized Challenges:** The AI assigns personalized math challenges based on student performance, providing targeted practice and feedback.

- **Engaging Games:** The platform offers engaging math games that motivate students to practice their math skills.

- **Progress Monitoring:** Teachers can monitor student progress through the platform's analytics dashboard, identifying areas where students need additional support.

Case Study: Prodigy Math in Elementary Schools

Context: An elementary school implemented Prodigy Math to support math instruction and practice.

Implementation: Students used Prodigy Math during their math class and for homework. The AI provided personalized math challenges and games.

Outcome: Students demonstrated increased motivation and engagement in math. Teachers reported that Prodigy Math helped students better understand math concepts and improve their skills.

7.3 Supporting English Language Learners with AI

English language learners (ELLs) often face unique challenges that require targeted support. AI can provide personalized language learning experiences that address ELLs' specific needs.

Language Learning Apps

AI-powered language learning apps can provide personalized content and practice for ELLs.

Example: Duolingo

Duolingo is an AI-driven language learning app that offers personalized lessons and practice for various languages, including English.

Implementation:

- **Personalized Lessons:** The AI assigns personalized lessons based on the learner's proficiency level, providing targeted practice and feedback.

- **Interactive Exercises:** The platform offers interactive exercises that cover vocabulary, grammar, listening, and speaking skills.

- **Progress Monitoring:** Learners can track their progress through the app's analytics dashboard, identifying areas where they need additional practice.

Case Study: Duolingo in Middle School ELL Programs

Context: A middle school implemented Duolingo to support ELLs in learning English.

Implementation: ELL students used Duolingo for daily language practice. The AI provided personalized lessons and interactive exercises.

Outcome: Students demonstrated significant improvement in English language proficiency. Teachers reported that Duolingo helped students build vocabulary and improve their speaking and listening skills.

Pronunciation and Conversation Practice

AI technologies can offer pronunciation guidance and simulate conversations for practice.

Example: Elsa Speak

Elsa Speak is an AI-powered app that helps learners improve their English pronunciation.

Implementation:

- **Pronunciation Practice:** The AI analyzes learners' pronunciation and provides real-time feedback and corrections.

- **Interactive Exercises:** The platform offers interactive exercises that cover common pronunciation challenges and practice sentences.

- **Conversation Practice:** Learners can practice conversations with the AI and receive feedback on pronunciation and fluency.

Case Study: Elsa Speak in High School ELL Programs

Context: A high school ELL program implemented Elsa Speak to help students improve their English pronunciation.

Implementation: Students used Elsa Speak for daily pronunciation practice. The AI provided real-time feedback and interactive exercises.

Outcome: Students demonstrated significant improvement in pronunciation and speaking confidence. Teachers reported that Elsa Speak helped students reduce their accents and improve their communication skills.

Cultural Context in Language Learning

AI tools can incorporate cultural context into language learning, making it more relatable and effective.

Example: Rosetta Stone

Rosetta Stone is an AI-driven language learning platform that integrates cultural context into its lessons.

Implementation:

- **Cultural Lessons:** The AI provides lessons that include cultural information and context, helping learners understand the language in real-life situations.

- **Interactive Exercises:** The platform offers interactive exercises that cover vocabulary, grammar, listening, and speaking skills.

- **Immersive Experience:** Learners are exposed to real-life scenarios and conversations, making the learning experience more immersive and compelling.

Case Study: Rosetta Stone in College ELL Programs

Context: A college ELL program implemented Rosetta Stone to support English language learning.

Implementation: ELL students used the Rosetta Stone for daily language practice. The AI provided personalized lessons and interactive exercises that included cultural context.

Outcome: Students demonstrated significant improvement in English language proficiency and cultural understanding. Teachers reported that Rosetta Stone helped students become more confident and effective communicators.

Real-Time Translation Tools

AI-driven real-time translation tools can aid comprehension and learning in the classroom.

Example: Google Translate

Google Translate is an AI-powered translation tool providing real-time text and speech translation.

Implementation:

- **Translation Assistance:** Students use Google Translate to translate text and speech in real-time, aiding comprehension and communication.

- **Classroom Support:** Teachers use Google Translate to provide instructions and explanations in the student's native language, ensuring understanding.

- **Interactive Learning:** The platform offers interactive translation exercises that help students practice and develop their language skills.

Case Study: Google Translate in High School ELL Programs

Context: A high school ELL program implemented Google Translate to support English language learning and comprehension.

Implementation: Students used Google Translate to translate text and speech in real-time, aiding comprehension and communication. Teachers used the platform to provide instructions and explanations in the student's native language.

Outcome: Students demonstrated improved comprehension and communication skills. Teachers reported that Google Translate helped students understand instructions and participate more actively in class.

7.4 AI for Special Education: Tools and Tactics

AI can provide customized learning experiences and support for students with special needs, making education more inclusive and effective.

Customized Learning Experiences

AI can create highly customized learning experiences for students with special needs, addressing their unique challenges and strengths.

Example: Kurzweil 3000

Kurzweil 3000 is an AI-powered assistive technology that provides reading, writing, and study support for students with learning disabilities.

Implementation:

- **Personalized Support:** The AI provides personalized reading and writing support, including text-to-speech, speech-to-text, and word prediction tools.

- **Interactive Exercises:** The platform offers interactive exercises that cover vocabulary, comprehension, and writing skills.

- **Progress Monitoring:** Teachers can monitor student progress through the platform's analytics dashboard, identifying areas where students need additional support.

Case Study: Kurzweil 3000 in Special Education Programs

Context: A unique education program implemented Kurzweil 3000 to support students with learning disabilities.

Implementation: Students used Kurzweil 3000 for reading and writing assignments. The AI provided personalized support and interactive exercises.

Outcome: Students demonstrated significant improvement in reading and writing skills. Teachers reported that Kurzweil 3000 helped students become more confident and independent learners.

Speech-to-Text for Communication

AI-driven speech-to-text technologies can assist students with communication challenges by converting spoken language into written text.

Example: Dragon NaturallySpeaking

Dragon NaturallySpeaking is an AI-powered speech recognition software that converts spoken language into written text.

Implementation:

- **Speech-to-Text:** Students use the software to dictate their thoughts and ideas, which are converted into written text in real-time.

- **Interactive Exercises:** The platform offers interactive exercises that help students practice and improve their speaking and writing skills.

- **Classroom Support:** Teachers use the software to support students with communication challenges, ensuring they can participate in class activities.

Case Study: Dragon NaturallySpeaking in Special Education Programs

Context: A unique education program implemented Dragon NaturallySpeaking to support students with communication challenges.

Implementation: Students used the software for writing assignments and classroom participation. The AI provided real-time speech-to-text conversion and interactive exercises.

Outcome: Students demonstrated improved communication and writing skills. Teachers reported that Dragon NaturallySpeaking helped students express their thoughts and ideas more effectively.

Behavioral Analysis and Support

AI tools can analyze behavioral patterns and support students with behavioral needs.

Example: Brain Power

Brain Power is an AI-powered platform that provides behavioral support for students with autism.

Implementation:

- **Behavioral Analysis:** The AI analyzes student behavior and provides insights and recommendations for support.

- **Interactive Exercises:** The platform offers interactive exercises that help students develop social and emotional skills.

- **Progress Monitoring:** Teachers and parents can monitor student progress through the platform's analytics dashboard, identifying areas where students need additional support.

Case Study: Brain Power in Special Education Programs

Context: A special education program implemented Brain Power to support students with autism.

Implementation: Students used Brain Power to practice daily social and emotional skills. The AI provided behavioral analysis and interactive exercises.

Outcome: Students demonstrated improved social and emotional skills. Teachers and parents reported that Brain Power helped students develop better self-regulation and communication skills.

Accessibility Tools

AI-powered accessibility tools can adapt content for students with visual, auditory, or physical disabilities, making education more inclusive.

Example: Microsoft Immersive Reader

Microsoft Immersive Reader is an AI-driven tool that enhances reading and writing accessibility for students with disabilities.

Implementation:

- **Customized Reading Options:** The AI provides customizable text options, such as font size, format, and background color, to accommodate different needs.

- **Text-to-Speech:** The platform offers text-to-speech functionality, allowing students to listen to written content.

- **Reading Comprehension:** The AI provides tools for improving reading comprehension, such as highlighting and dictionary support.

Case Study: Microsoft Immersive Reader in Special Education Programs

Context: A special education program implemented Microsoft Immersive Reader to support students with visual and reading disabilities.

Implementation: Students used Microsoft Immersive Reader for reading assignments and classroom activities. The AI provided customizable text options and text-to-speech functionality.

Outcome: Students demonstrated improved reading comprehension and engagement. Teachers reported that Microsoft Immersive Reader made reading more accessible and enjoyable for students with disabilities.

Conclusion

AI has the potential to transform education by addressing diverse learning needs and providing personalized support to students with varying abilities and challenges. By leveraging AI-powered tools and technologies, educators can create more inclusive and effective learning environments that cater to the unique needs of all students. The examples and case studies in this chapter illustrate the diverse ways AI can be integrated into education to support struggling readers, customize math instruction, assist English language learners, and provide tools for special education. As educators continue exploring and adopting AI technologies, they can unlock new possibilities for teaching and learning, benefiting students and enriching the educational experience.

Chapter 8: AI for Professional Development

8.1 Using AI to Personalize Professional Learning for Educators

AI technology can revolutionize professional development for educators, empowering them with personalized learning experiences tailored to their individual needs and goals. This chapter delves into how AI can enhance professional development, providing specific examples and case studies.

Tailored Professional Development

AI platforms play a crucial role in professional development by identifying educators' strengths and development areas, paving the way for personalized improvement paths.

Example: Coursera for Educators

Coursera offers many online courses, several of which are designed specifically for educators. Using AI, Coursera can recommend courses that match an educator's interests and professional development needs.

Implementation:

- **Initial Assessment:** Educators complete an initial assessment to identify their skills and areas for improvement.

- **Personalized Recommendations:** The AI analyzes the assessment data and recommends courses that align with the educator's goals and professional development needs.

- **Progress Tracking:** The platform tracks progress and provides feedback, allowing educators to see their growth over time.

Case Study: Coursera for Educators in a School District

Context: A school district implemented Coursera for Educators to provide personalized professional development for its teachers.

Implementation: Teachers completed an initial assessment, and the AI recommended courses in areas such as classroom management, differentiated instruction, and technology integration. Teachers enrolled in the recommended courses and completed them at their own pace.

Outcome: Teachers reported increased confidence and competence in their teaching practices. The personalized nature of the courses ensured that professional development was relevant and impactful, leading to improved classroom outcomes.

AI in Classroom Observation

AI tools can assist with classroom observation by providing real-time, objective feedback and guidance regarding areas for improvement.

Example: TeachFX

TeachFX is an AI-powered tool that analyzes classroom audio recordings to provide insights into teacher-student interactions.

Implementation:

- **Audio Recording:** Teachers record their lessons using the TeachFX app.

- **AI Analysis:** The AI analyzes the audio data to identify teacher and student talk time patterns, question types, and engagement levels.

- **Feedback and Recommendations:** The platform provides actionable recommendations to help teachers improve their instructional practices.

Case Study: TeachFX in Middle Schools

Context: A middle school implemented TeachFX to support professional development and improve instructional practices.

Implementation: Teachers recorded their lessons and used TeachFX to analyze the
data. The AI provided feedback on questioning techniques, student engagement, and teacher talk time.

Outcome: Teachers reported increased awareness of their instructional practices and made data-driven adjustments to their teaching. The feedback from TeachFX helped them create more interactive and engaging lessons.

Continuous Learning and Improvement

AI tools can support continuous learning and improvement by identifying areas for growth and recommending resources.

Example: Edthena

Edthena is an AI-powered platform that uses video analysis to support teacher reflection and professional development.
   Implementation:

- **Video Recording:** Teachers record and upload their lessons to the Edthena platform.

AI Analysis: The AI analyzes the videos and provides feedback on teaching aspects, such as instructional strategies, classroom management, and student engagement.

- **Personalized Recommendations:** Based on the analysis, the platform recommends resources and professional development opportunities to address identified areas for growth.

Case Study: Edthena in High Schools

Context: A high school implemented Edthena to enhance professional development and support teacher growth.

Implementation: Teachers recorded their lessons and uploaded them to Edthena. The AI provided feedback and recommendations for professional development.

Outcome: Teachers reported increased reflection on their teaching practices and a greater willingness to try new strategies. The personalized recommendations helped them access relevant resources and improve their instructional techniques.

Self-Assessment Tools

AI-powered self-assessment tools can help educators evaluate their skills and track their progress.

Example: Skillshare for Educators

Skillshare offers many online courses focused on professional development for educators. The platform uses AI to help educators assess their skills and track their progress.

Implementation:

- **Initial Assessment:** Educators complete a self-assessment to identify their current skills and development areas.

- **Personalized Recommendations:** The AI recommends courses based on the self-assessment results.

- **Progress Tracking:** The platform tracks progress and provides feedback, allowing educators to see their growth over time.

Case Study: Skillshare for Educators in a School District

Context: A school district implemented Skillshare for Educators to provide personalized professional development for its teachers.

Implementation: Teachers completed an initial self-assessment, and the AI recommended courses in classroom management, differentiated instruction, and technology integration. Teachers enrolled in the recommended courses and completed them at their own pace.

Outcome: Teachers reported a significant boost in confidence and competence in their teaching practices. The personalized nature of the courses, guided by AI, ensured that professional development was not just relevant, but also impactful, leading to improved classroom outcomes.

8.2 Collaborative AI Learning Platforms for Teachers

Collaborative AI learning platforms are not just tools, but supportive communities that can facilitate professional learning and peer-to-peer collaboration among educators.

Online Professional Learning Communities

AI-powered platforms can create dynamic online communities where educators can exchange experiences and learn from each other.

Example: Microsoft Teams for Education

Microsoft Teams for Education provides a collaborative workspace where educators can share resources, participate in discussions, and work on projects together.

Implementation:

- **Team Creation:** Educators create teams based on grade level, subject area, or professional development goals.

- **Resource Sharing:** The AI recommends resources and facilitates sharing within the team.

- **Discussion and Collaboration:** Educators participate in discussions, collaborate on projects, and provide feedback to each other.

Case Study: Microsoft Teams for Education in a School District

Context: A school district implemented Microsoft Teams for Education to support collaborative professional development.

Implementation: Teachers created teams based on their professional development goals. The AI recommended resources and facilitated discussions and collaboration within the teams.

Outcome: Teachers reported increased collaboration and sharing of best practices. The platform helped create a supportive professional learning community, improving instructional strategies and student outcomes.

Shared Resource Libraries

AI-powered platforms can curate and recommend teaching resources based on educators' interests and teaching contexts.

Example: Teachers Pay Teachers

Teachers Pay Teachers is a digital marketplace where educators can buy, sell, and share original teaching resources. The platform uses AI to recommend resources based on user preferences and needs.

Implementation:

- **Resource Curation:** The AI curates resources based on the educator's teaching context, interests, and past purchases.

- **Personalized Recommendations:** Educators receive personalized recommendations for resources that align with their professional development goals and classroom needs.

- **Community Engagement:** Educators can share their resources and collaborate with others on the platform.

Case Study: Teachers Pay Teachers in Middle Schools

Context: A middle school used Teachers Pay Teachers to support professional development and resource sharing.

Implementation: Teachers created profiles and used the AI to receive personalized resource recommendations. They also shared their resources and collaborated with others on the platform.

Outcome: Teachers reported finding high-quality resources that met their specific needs. The platform facilitated collaboration and sharing of best practices, leading to improved instructional strategies and student outcomes.

Mentorship Matching

AI can connect less experienced teachers with mentors, enhancing professional development through shared knowledge and support.

Example: MentorMatch

MentorMatch is an AI-powered platform that connects educators with mentors based on their professional development needs and goals.

Implementation:

- **Profile Creation:** Educators create profiles that include their professional development goals, teaching experience, and areas of interest.

Mentor Matching: The AI matches new teachers with more experienced mentors who possess the relevant expertise and can provide guidance and support.

- **Ongoing Support:** The platform facilitates ongoing communication and collaboration between mentors and mentees.

Case Study: MentorMatch in High Schools

Context: A high school implemented MentorMatch to support professional development and mentorship for new teachers.

Implementation: New teachers created profiles and were matched with experienced mentors. The AI facilitated ongoing communication and collaboration between mentors and mentees.

Outcome: New teachers reported feeling more supported and confident in their teaching practices. The mentorship program helped them navigate challenges and improve their instructional strategies.

Collaborative Project Tools

AI tools can support collaborative projects and research among educators across different locations.

Example: Google Workspace for Education

Google Workspace for Education provides a suite of tools that facilitate collaboration on projects and research.

Implementation:

- **Project Management:** Educators use shared Google Drive tools like Docs, Sheets, and Slides to collaborate on documents and research.

- **Real-Time Collaboration:** The AI facilitates real-time collaboration, allowing educators to collaborate seamlessly, regardless of location.

- **Resource Sharing:** Educators can share resources and provide feedback using the platform's collaboration tools.

Case Study: Google Workspace for Education in a School District

Context: A school district used Google Workspace for Education to support collaborative professional development projects.

Implementation: Teachers collaborated on projects and research using Google Docs, Sheets, and Slides. The AI facilitated real-time collaboration and resource sharing.

Outcome: Teachers reported increased collaboration and sharing of best practices. The platform helped them work together on professional development projects, leading to improved instructional strategies and student outcomes.

8.3 Keeping Up with AI Trends: Resources for Educators

Educators must stay informed about the latest AI trends and developments. This section explores resources that can help educators stay current.

Curated AI News and Updates

AI-powered platforms can curate news and updates on the latest AI and education technology developments.

Example: Feedly

Feedly is an AI-driven news aggregator that curates news and updates based on user interests.

Implementation:

- **Interest Profiles:** Educators create profiles that include their interests and areas of focus.

- **Curated Content:** The AI curates news and updates on AI and education technology, delivering personalized content to educators.

- **Continuous Learning:** Educators can stay informed about the latest trends and developments by regularly checking their curated feed.

Case Study: Feedly for Educators

Context: A group of educators used Feedly to stay informed about AI and education technology trends.

Implementation: Educators created profiles and received curated news and updates on AI and education technology.

Outcome: Educators reported feeling more informed and up-to-date on the latest trends and developments. The curated content helped them integrate new ideas and technologies into their teaching practices.

Webinars and Online Courses

Online learning opportunities focused on AI can help educators develop practical, classroom-applicable skills.

Example: edX

edX offers many online courses and webinars on AI and education technology.

Implementation:

- **Course Enrollment:** Educators enroll in courses and webinars on AI and education technology.

- **Interactive Learning:** Courses include interactive lessons, quizzes, and discussions to reinforce learning.

- **Practical Applications:** Educators apply their knowledge through projects and assignments relevant to their teaching context.

Case Study: edX in Professional Development

Context: A school district encouraged teachers to enroll in edX courses on AI and education technology.

Implementation: Teachers enrolled in courses and webinars on AI and education technology. They completed interactive lessons, quizzes, and projects.

Outcome: Teachers reported increased knowledge and confidence in integrating AI and education technology into their teaching. The practical applications helped them develop relevant skills and improve their instructional strategies.

AI Tool Reviews and Recommendations

Platforms that review and recommend AI tools for education can help educators make informed decisions.

Example: Common Sense Education

Common Sense Education provides reviews and recommendations for AI tools and education technology.

Implementation:

- **Tool Reviews:** Educators access reviews of AI tools and education technology, including strengths, weaknesses, and classroom applications.

- **Personalized Recommendations:** The AI provides personalized recommendations based on the educator's teaching context and needs.

- **Resource Sharing:** Educators can share their reviews and recommendations with the community.

Case Study: Common Sense Education in Middle Schools

Context: A middle school used Common Sense Education to identify and implement AI tools and education technology.

Implementation: Teachers accessed reviews and received personalized recommendations for AI tools and education technology. They shared their own experiences and suggestions with the community.

Outcome: Teachers reported finding high-quality AI tools that met their specific needs. The platform facilitated informed decision-making and sharing of best practices, leading to improved instructional strategies and student outcomes.

Networking with AI Experts

Participating in conferences and forums where educators can network with AI researchers and practitioners can provide valuable insights and opportunities for collaboration.

Example: The International Society for Technology in Education (ISTE) Conference

ISTE Conference is an annual event that brings educators, researchers, and technology experts together to discuss the latest developments in education technology.

Implementation:

Conference Participation: Educators attend the ISTE Conference to learn about the recent developments in AI and advancements in education technology.

- **Networking Opportunities:** The conference allows educators to network with AI researchers and practitioners, sharing insights and best practices.

- **Workshops and Sessions:** The conference includes workshops and sessions on AI and education technology, providing practical, classroom-applicable knowledge.

Case Study: ISTE Conference in Professional Development

Context: A group of educators attended the ISTE Conference to learn about AI and education technology.

Implementation: Educators participated in workshops and sessions on AI and education technology. They networked with AI researchers and practitioners, sharing insights and best practices.

Outcome: Educators reported increased knowledge and confidence in integrating AI and education technology into their teaching. The networking opportunities provided valuable connections and insights, leading to improved instructional strategies and student outcomes.

8.4 Establishing a Professional Learning Network around AI in Education

Building a Professional Learning Network (PLN) focused on AI in education can provide ongoing support and collaboration for educators.

Building a PLN

This section will guide educators in building a Professional Learning Network (PLN) focused on AI in education, including steps to get started and platforms to use.

Example: Twitter for Educators

Twitter, now known as X, is a social media platform that can build a PLN focused on AI in education.

Implementation:

- **Creating an Account:** Educators create Twitter accounts and follow experts and organizations focused on AI in education.

- **Engaging in Discussions:** Educators participate in discussions, share resources, and collaborate with others in their PLN.

- **Using Hashtags:** Educators use hashtags such as #AIinEducation and #EdTech to find and contribute to discussions and resources.

Case Study: Twitter for Educators in a School District

Context: A school district encouraged teachers to use Twitter to build a PLN focused on AI in education.

Implementation: Teachers created Twitter accounts and followed experts and organizations focused on AI in education. They participated in discussions, shared resources, and collaborated with others in their PLN.

Outcome: Teachers reported increased collaboration and sharing of best practices. The PLN provided ongoing support and access to valuable resources, leading to improved instructional strategies and student outcomes.

Sharing Best Practices

Encourage sharing AI integration strategies and successes within the PLN to foster collective learning and improvement.

Example: Edmodo

Edmodo is a social learning platform that allows educators to share resources, collaborate on projects, and engage in discussions.

Implementation:

- **Creating Groups:** Educators create groups focused on AI in education to share resources and best practices.

- **Collaborative Projects:** Educators collaborate on projects and research, sharing their findings and insights with the group.

- **Ongoing Discussions:** Educators engage in ongoing discussions, asking questions and providing feedback to each other.

Case Study: Edmodo in Professional Development

Context: A group of educators used Edmodo to create a PLN focused on AI in education.

Implementation: Educators created groups focused on AI in education, sharing resources and best practices. They collaborated on projects and engaged in ongoing discussions.

Outcome: Educators reported increased collaboration and sharing of best practices. The PLN provided ongoing support and access to valuable resources, leading to improved instructional strategies and student outcomes.

Collaborative Problem-Solving

Highlight how a PLN can serve as a support system for troubleshooting and problem-solving AI-related challenges in the classroom.

Example: Google Groups

Google Groups is a platform that allows educators to create and participate in discussion groups focused on specific topics.

Implementation:

- **Creating Groups:** Educators create Google Groups focused on AI in education, where they can ask questions and share solutions.

- **Collaborative Problem-Solving:** Educators collaborate to troubleshoot and solve AI-related challenges in the classroom.

- **Resource Sharing:** Educators share resources and best practices within the group.

Case Study: Google Groups in Professional Development

Context: A group of educators used Google Groups to create a PLN focused on AI in education.

Implementation: Educators created Google Groups focused on AI in education, where they asked questions and shared solutions. They collaborated to troubleshoot and solve AI-related challenges in the classroom.

Outcome: Educators reported increased collaboration and problem-solving. The PLN provided ongoing support and access to valuable resources, leading to improved instructional strategies and student outcomes.

Global Perspectives

Discuss the benefits of including diverse, global perspectives in a PLN to broaden understanding of AI's impact on education worldwide.

Example: LinkedIn

LinkedIn is a professional networking platform that allows educators to connect with peers and experts worldwide.

Implementation:

- **Creating Profiles:** Educators create LinkedIn profiles and connect with peers and experts focused on AI in education.

- **Engaging in Discussions:** Educators participate in discussions and share resources with their global network.

- **Collaborative Projects:** Educators collaborate on projects and research with peers and experts worldwide.

Case Study: LinkedIn in Professional Development

Context: A group of educators used LinkedIn to create a global PLN focused on AI in education.

Implementation: Educators created LinkedIn profiles and connected with peers and experts worldwide. They participated in discussions, shared resources, and collaborated on projects.

Outcome: Educators reported increased access to diverse perspectives and best practices. The global PLN provided valuable insights and resources, leading to improved instructional strategies and student outcomes.

Conclusion

AI has the potential to revolutionize professional development for educators by offering personalized learning experiences, facilitating collaboration, and providing access to valuable resources and insights. By leveraging AI-powered tools and platforms, educators can create dynamic and supportive professional learning environments, while educator communities can stay informed about the latest trends and advancements in AI. They can continuously improve their instructional practices using various professional development tools. The examples and case studies in this chapter illustrate the diverse ways AI can be integrated into professional development, benefiting educators and enhancing the educational experience for their students.

11

Chapter 9: Understanding Machine Learning and Natural Language Processing

9.1 Machine Learning Demystified for Educators

Machine Learning (ML) is an element of AI that empowers educators by training algorithms to recognize patterns in data, make predictions, and improve over time through experience. Understanding ML can unlock numerous possibilities for educators to enhance teaching and learning, giving them the confidence to embrace these technologies.

Simplifying Machine Learning Concepts

Definition and Types of Machine Learning:

- **Supervised Learning:** Supervised learning algorithms are programmed using labeled data, facilitating accurate output. This type of learning is helpful for classification (e.g., spam detection) and regression tasks (e.g., predicting student grades).

- **Unsupervised Learning:** The algorithm is given input data without labeled responses and must find patterns or groupings independently. Examples include clustering students based on learning styles.

- **Reinforcement Learning:** The algorithm learns by interacting with an environment and receiving feedback through rewards or penalties. This approach can be used for adaptive learning systems, where the software adjusts based on student interactions.

Example Application in Education:

- **Predicting Student Performance:** Supervised learning algorithms can predict student performance by analyzing integrated historical data, such as student grades, engagement, and attendance. This allows educators to identify at-risk students early and intervene with targeted support.

Tools and Resources for Educators

Educators do not need to be experts in coding to leverage machine learning tools. Several user-friendly platforms are available to make adopting AI tools practical and comfortable for educators.

Google's Teachable Machine is an accessible platform where teachers and students can train simple ML models using their data. This platform allows you to easily create models that recognize and respond to specific gestures, sounds, or images, making it an excellent tool for interactive learning.

- **Microsoft Azure Machine Learning:** Offers drag-and-drop interfaces to build ML models without writing code.

- **IBM Watson:** Provides tools for developing ML models and integrating them into educational applications.

9.2 The Role of Natural Language Processing in Educational AI

Natural Language Processing (NLP) is an AI field focused on interactions between computer networks and human languages. NLP enables machines to understand, interpret, and respond to human language in a valuable way.

Basics of Natural Language Processing

Core Components of NLP:

- **Tokenization:** Breaking down text into smaller units (tokens), such as words or sentences.

- **Part-of-Speech Tagging:** Identifying the grammatical elements of speech (nouns, verbs, adjectives) within the text.

- **Named Entity Recognition (NER):** Detecting and classifying entities (e.g., names, dates, locations) within the text.

- **Sentiment Analysis:** Determining the sentiment or emotion expressed in the text.

Example Application in Education:

- **Automated Essay Scoring**: NLP algorithms can evaluate the quality of student essays by analyzing grammar, coherence, and argumentation. Tools like **Grammarly** and **Criterion Online Writing Evaluation Service** provide real-time writing feedback and grading assistance.

Tools and Resources for Educators

Several NLP tools can be integrated into classroom activities to enhance learning:

- **Grammarly:** Offers grammar, spelling, and style suggestions to help students improve their writing.

- **Google Translate:** Assists language learners by providing translations and pronunciation guides.

- **Turnitin:** Uses NLP to detect plagiarism and check the originality of student submissions.

9.3 Advanced AI: Neural Networks and Their Educational Implications

Neural Networks are the essence of deep learning, a subset of machine learning. Developed based on the structure and function of the human brain, neural networks consist of interconnected layers of nodes (neurons) that process data.

Understanding Neural Networks

Components of Neural Networks:

- **Input Layer:** Receives the raw input data.

Hidden Layers: These perform computations and extract features from the data. The number of hidden layers varies depending on the complexity of the problem.

- **Output Layer:** Produces the final output, such as a classification label or a numerical value.

Example Application in Education:

- **Adaptive Learning Platforms:** Neural networks can power adaptive learning platforms like **Knewton** and **DreamBox**, which provide personalized learning experiences by continuously analyzing student interactions and adjusting content accordingly.

Preparing for Advanced AI

Educators can take steps to prepare themselves and their students for the integration of advanced AI technologies:

- **Professional Development:** Participate in workshops and online courses on AI and machine learning. Platforms like **edX** and **Coursera** offer specialized courses for educators.

- **Collaborative Projects:** Engage students in collaborative projects that involve building simple neural network models. Tools like **TensorFlow** and **Keras** provide accessible frameworks for educational use.

9.4 Developing AI Literacy: Essential Skills for Educators

AI Literacy is not just a buzzword; it's a crucial skill set for educators and students alike. Understanding the basics of AI, its applications, and its implications is a responsibility for educators. It's a way to equip students with the skills they need to navigate a future where AI is not just a possibility but a certainty.

Defining AI Literacy

AI literacy encompasses several key areas:

- **Understanding AI Concepts:** Familiarity with fundamental AI concepts, including machine learning, gamification, and natural language processing.

- **Ethical Considerations:** Remember to fully address ethical issues related to AI, such as data privacy, bias, and the impact on employment.

- **Practical Applications**: Knowledge of how AI can enhance learning and teaching in educational settings.

Example Application in Education:

- **AI Workshops:** Schools can organize AI literacy workshops for teachers and students, covering basic AI principles, practical applications, and ethical considerations.

Building AI Literacy

Educators can build AI literacy through various strategies:

- **Online Courses and Certifications:** Sign up for online courses and certification programs on AI. Platforms like **Udacity** and **IBM AI Education** offer courses tailored to educators.

- **Peer Learning:** Create a community of practice where educators can share resources, experiences, and best practices related to AI integration.

- **Student Projects:** Encourage students to explore AI through hands-on projects like creating chatbots or analyzing data with machine learning tools.

Integrating AI Literacy into Curriculum

Example Curriculum Integration:

- **Computer Science Classes**: Incorporate AI modules into computer science curriculums, teaching students about machine learning algorithms and neural networks.

- **Cross-Disciplinary Projects**: Design cross-disciplinary projects that involve AI, such as using NLP to analyze literary texts or applying machine learning to predict climate change impacts in science classes.

Conclusion

Understanding machine learning and natural language processing is essential for educators to leverage AI effectively in their teaching practices. By demystifying these concepts, exploring practical applications, and building AI literacy, educators can enhance student learning experiences and prepare them for a future where AI plays a pivotal role. AI skills will be critical to student success in the future, and those equipped with AI literacy will be competitive in the job market.

12

Chapter 10: Preparing for the Future of AI in Education

Integrating artificial intelligence (AI) in education promises a transformative impact on teaching and learning. This chapter explores how educators, administrators, and policymakers can prepare for the future of AI-driven personalized learning, address ethical considerations, adapt to educators' evolving roles, and embrace the global classroom concept.

10.1 The Future of AI-Driven Personalized Learning

AI-driven personalized learning tailors educational experiences to individual student needs, preferences, and learning paces. This approach enhances student engagement, improves learning outcomes, and helps close achievement gaps. As AI technology advances, it is crucial to follow emerging trends and prepare for the future of personalized learning.

Emerging Trends in Personalization

Adaptive Learning Technologies

Adaptive learning platforms use AI to adjust the difficulty and pace of lessons based on student performance.

Example: Knewton Alta

Knewton Alta is an AI platform that offers personalized coursework in various subjects, including math and science, through adaptive learning.

Implementation:

Initial Assessment: Students complete an initial assessment to determine their proficiency levels.

Personalized Coursework: The AI assigns personalized coursework that adapts in real-time based on student performance, providing targeted practice and feedback.

Progress Monitoring: Teachers can monitor student progress through the platform's analytics dashboard, identifying areas where students need additional support.

Case Study: Knewton Alta in College Math Courses

Context: A college math department implemented Knewton Alta to support students in introductory math courses.

Implementation: Students used Knewton Alta for coursework and homework. The AI provided personalized lessons and real-time feedback.

Outcome: Students demonstrated significant improvement in math proficiency and course completion rates. Teachers reported that Knewton Alta helped them effectively identify and address individual learning needs.

Predictive Analytics

Predictive analytics use AI to analyze historical and current data to predict future student outcomes and identify at-risk students.

Example: Blackboard Predict

Blackboard Predict uses predictive analytics to forecast student outcomes and suggest interventions to improve retention and success.

Implementation:

Data Collection: The platform collects student performance, attendance, and engagement data.

Predictive Models: The AI analyzes the data to identify patterns and predict future outcomes.

Interventions: Teachers receive alerts about at-risk students and suggestions for targeted interventions.

Case Study: Blackboard Predict in a University

Context: A university used Blackboard Predict to identify at-risk students in extensive introductory courses.

Implementation: Professors monitored student performance and engagement data through Blackboard Predict. The AI provided alerts and intervention suggestions for at-risk students.

Outcome: The university implemented targeted interventions for at-risk students, such as tutoring and academic support. As a result, student retention and success rates improved significantly.

Barriers and Solutions

Barrier: Data Privacy Concerns

Student data collection and use for personalized learning generates data privacy and security concerns.

Solution: Robust Data Privacy Policies

Educational institutions must implement robust data privacy policies and practices to protect student data. Compliance demands following relevant laws and regulations (e.g., FERPA and GDPR) and transparent data usage communication with students and parents.

Barrier: Resource Constraints

Implementing AI-driven personalized learning requires significant resources, including technology infrastructure and professional development for educators.

Solution: Strategic Investment and Partnerships

Schools and districts can seek strategic investments and partnerships to support the implementation of AI-driven personalized learning. These may include grants, private organization funding, and collaborations with edtech companies.

10.2 Ethical AI: Future Considerations for Educators

As AI becomes more prevalent in education, concerns must be addressed to ensure that AI is used equitably and ethically.

Evolving Ethical Standards

Transparency and Accountability

AI systems in education must be transparent and accountable. This includes clear documentation of how AI algorithms work, what data they use, and how decisions are made.

Example: Transparent AI Algorithms

Google AI Principles emphasize transparency and accountability in AI development and deployment.

Implementation:

Clear Documentation: AI developers provide clear documentation on how algorithms work and what data they use.

Ethical Guidelines: Institutions establish ethical guidelines for using AI in education, ensuring transparency and accountability.

Case Study: Google AI Principles in Education

Context: An educational technology company adopted Google AI Principles to guide the development of its AI-driven products.

Implementation: The company provided clear documentation on its AI algorithms and established ethical guidelines for their use in educational settings.

Outcome: The company's commitment to transparency and accountability built trust with educators and students, promoting responsible AI use.

Bias and Fairness

AI tools must be built and implemented to minimize bias and ensure fairness.

Example: IBM Watson for Education

IBM Watson for Education focuses on reducing bias and ensuring fairness in AI-driven educational tools.

Implementation:

Bias Mitigation: The AI development team identifies and mitigates potential biases in data and algorithms.

Fairness Audits: Regular fairness audits ensure equitable outcomes for all students.

Case Study: IBM Watson for Education

Context: IBM Watson for Education implemented bias mitigation strategies to ensure fairness in its AI tools.

Implementation: The team conducted fairness audits and adjusted algorithms and data sources to minimize bias.

Outcome: The AI tools provided more equitable outcomes for diverse student populations, promoting fairness in education.

Global Perspectives on Ethical AI

Different countries have varying perspectives on ethical AI in education. Understanding these perspectives can inform best practices and foster international collaboration.

Example: European Union's AI Ethics Guidelines

The European Union has established AI ethics guidelines emphasizing human oversight, transparency, and fairness.

Implementation:

Ethics Framework: Educational institutions in the EU implement the ethics framework to guide the use of AI in education.

Collaboration: Institutions collaborate with international partners to share best practices and promote ethical AI use globally.

Case Study: European Union's AI Ethics Guidelines in Education

Context: An EU-based educational institution adopted the EU's AI ethics guidelines to guide its use of AI in education.

Implementation: The institution implemented the ethics framework and exchanged best practices with international partners.

Outcome: The institution promoted ethical AI use and contributed to the global conversation on AI ethics in education.

10.3 The Evolving Role of the Educator in the AI-Enhanced Classroom

The integration of AI in education is changing the role of educators. While AI can automate specific tasks, educators remain essential for providing human connection, guidance, and support.

Changing Roles

From Instructor to Facilitator

AI can automate routine tasks, allowing educators to focus on facilitating learning and providing personalized support.

Example: AI-Assisted Grading

Gradescope uses AI to assist with grading assignments and exams, freeing teachers to focus on instruction and support.

Implementation:

Automated Grading: The AI grades assignments based on predefined criteria.

Personalized Support: Teachers use the time saved to provide personalized support and feedback to students.

Case Study: Gradescope in Higher Education

Context: A university used Gradescope to assist with grading in large classes.

Implementation: Professors used Gradescope for automated grading and focused on providing personalized support to students.

Outcome: Professors reported increased efficiency and improved student engagement. The AI-assisted grading allowed them to spend more time on instruction and support.

Skill Development

Educators must develop new skills to integrate AI into their teaching practices effectively.

Example: Professional Development Programs

edX offers professional development programs focused on AI and education technology for educators.

Implementation:

Course Enrollment: Educators enroll in professional development courses on AI and education technology.

Interactive Learning: Courses include interactive lessons, quizzes, and projects to reinforce learning.

Skill Application: Educators apply their new skills in the classroom, integrating AI tools into their teaching practices.

Case Study: edX Professional Development Programs

Context: A school district encouraged teachers to enroll in edX courses on AI and education technology.

Implementation: Teachers enrolled in courses and completed interactive lessons, quizzes, and projects.

Outcome: Teachers reported increased confidence and competence in integrating AI into their teaching. The professional development programs helped them develop relevant skills and improve their instructional strategies.

Professional Development Pathways

Ongoing Learning

Continuous professional development is essential for educators to keep up with the trends and advancements in technology.

Example: ISTE Standards for Educators

The International Society for Technology in Education (ISTE) provides standards and resources for ongoing professional development in technology integration.

Implementation:

Standards Alignment: Schools align their professional development programs with ISTE standards.

Resource Access: Educators access ISTE resources, including courses, webinars, and workshops, to stay updated on AI and education technology.

Community Engagement: Educators participate in the ISTE community, sharing best practices and collaborating with peers.

Case Study: ISTE Standards for Educators

Context: A school district aligned its professional development programs with ISTE standards.

Implementation: Teachers accessed ISTE resources and participated in courses, webinars, and workshops. They engaged with the ISTE community to share best practices and collaborate with peers.

Outcome: Teachers reported increased knowledge and confidence in integrating AI and education technology. The ongoing professional development helped them stay updated on the latest trends and advancements.

Collaborative Learning

Professional learning communities can support collaborative learning and sharing of best practices among educators.

Example: Microsoft Teams for Education

Microsoft Teams for Education provides a collaborative workspace for professional learning communities.

Implementation:

Team Creation: Educators create teams based on grade level, subject area, or professional development goals.

Resource Sharing: The AI recommends resources and facilitates sharing within the team.

Discussion and Collaboration: Educators participate in discussions, collaborate on projects, and provide feedback to each other.

Case Study: Microsoft Teams for Education in a School District

Context: A school district used Microsoft Teams for Education to support collaborative professional development.

Implementation: Teachers created teams based on their professional development goals. The AI recommended resources and facilitated discussions and collaboration within the teams.

Outcome: Teachers reported increased collaboration and sharing of best practices. The platform helped create a supportive professional learning community, improving instructional strategies and student outcomes.

10.4 AI and the Global Classroom: Bridging Educational Gaps

AI can create a global classroom where students from diverse backgrounds can access quality education and collaborate across borders.

The Global Classroom Concept

Virtual Learning Environments

AI-powered virtual learning environments can connect students from different regions, providing access to diverse perspectives and resources.

Example: Google Classroom

Google Classroom is a virtual learning environment facilitating collaboration and resource sharing among students and teachers.

Implementation:

Classroom Creation: Teachers create virtual classrooms and invite students from different regions.

Resource Sharing: The AI recommends resources and facilitates sharing within the classroom.

Collaborative Projects: Students collaborate on projects and assignments, sharing their perspectives and learning from each other.

Case Study: Google Classroom in a Global Learning Initiative

Context: An international learning initiative used Google Classroom to connect students from different countries.

Implementation: Teachers created virtual classrooms and facilitated collaborative projects and discussions among students from different regions.

Outcome: Students reported increased cultural understanding and engagement. The virtual learning environment provided access to diverse perspectives and resources, enriching the educational experience.

Bridging Educational Gaps

Equitable Access to Education

AI can help bridge educational gaps by providing equitable access to quality education for students in underserved regions.

Example: Khan Academy

Khan Academy offers free, AI-powered educational resources to students worldwide.

Implementation:

Resource Access: Students access Khan Academy's free educational resources, including videos, exercises, and quizzes.

Personalized Learning: The AI provides personalized learning paths based on student performance.

Progress Monitoring: Teachers and parents can monitor student progress through the platform's analytics dashboard.

Case Study: Khan Academy in Underserved Regions

Context: Schools in underserved regions implemented Khan Academy to provide access to quality education.

Implementation: Students accessed Khan Academy's educational resources and personalized learning paths, and teachers and parents monitored student progress through the platform.

Outcome: Students demonstrated significant improvement in academic performance. The platform provided equitable access to quality education, helping bridge educational gaps.

Cultural Exchange and Collaboration

AI-driven projects can promote cultural exchange and collaboration among students from different backgrounds.

Example: ePals

ePals is an AI-powered platform that connects classrooms worldwide for collaborative projects and cultural exchange.

Implementation:

Classroom Connections: Teachers connect their classrooms with others from different countries.

Collaborative Projects: Students work on collaborative projects, sharing their perspectives and learning from each other.

Cultural Exchange: The platform facilitates cultural exchange through discussions, videos, and interactive activities.

Case Study: ePals in a Cultural Exchange Program

Context: A cultural exchange program used ePals to connect classrooms from different countries.

Implementation: Teachers connected their classrooms and facilitated collaborative projects and cultural exchange activities.

Outcome: Students reported increased cultural understanding and engagement. The platform allowed students to learn from each other and develop global perspectives.

Challenges and Opportunities

Challenge: Digital Divide

With unequal technology and internet connectivity access, the digital divide remains a significant challenge.

Opportunity: Strategic Investments

Strategic investments in technology infrastructure and digital literacy programs can help bridge the digital divide.

Example: One Laptop per Child

One Laptop per Child (OLPC) is an initiative that provides affordable laptops to students in underserved regions.

Implementation:

Laptop Distribution: The initiative distributes affordable laptops to students in underserved regions.

Digital Literacy Programs: The initiative offers digital literacy programs to help students and teachers use the technology effectively.

Case Study: One Laptop per Child in a Rural Community

Context: The OLPC initiative was implemented in a rural community to provide access to technology and digital literacy.

Implementation: Students received affordable laptops, and digital literacy programs were offered to help them and their teachers use the technology effectively.

Outcome: Students demonstrated increased digital literacy and engagement. The initiative helped bridge the digital divide and provided access to quality education.

Challenge: Language Barriers

Language barriers can limit access to global educational resources and collaboration opportunities.

Opportunity: Multilingual AI Tools

Multilingual AI tools can help overcome language barriers, providing real-time translation and language support.

Example: Microsoft Translator

Microsoft Translator is an AI-powered translation tool that provides real-time translation and language support.

Implementation:

Classroom Use: Teachers use Microsoft Translator to provide instructions and explanations in students' native languages.

Student Use: Students use the platform to translate texts and improve their understanding of different languages.

Interactive Learning: The platform offers interactive translation exercises that help learners practice and develop their language skills.

Case Study: Microsoft Translator in a Multilingual Classroom

Context: A multilingual classroom used Microsoft Translator to support ELL students.

Implementation: Teachers translated instructions and materials using

Microsoft Translator. Students used the platform for reading assignments and vocabulary practice.

Outcome: Students demonstrated improved comprehension and language skills. Teachers reported that Microsoft Translator helped bridge language gaps and facilitate better communication.

Conclusion

The future of AI in education offers tremendous potential for enhancing personalized learning, addressing ethical considerations, evolving the role of educators, and creating a global classroom. Educators, administrators, and policymakers can leverage AI to improve educational outcomes and provide equitable access to quality education for all students by understanding and preparing for these developments. The examples and case studies in this chapter illustrate the diverse ways AI can be integrated into education, benefiting students and enriching the educational experience. As AI technology evolves, the education community must stay informed, collaborate, and adapt to ensure that AI is used responsibly and effectively in pursuing educational excellence.

13

Chapter 11: Building an AI-Ready Classroom

Creating an AI-ready classroom is not just about applying AI tools and technology, but also about empowering educators and school administrators to lead the transformation. This chapter explores the infrastructure and resources needed for AI adoption, strategies to overcome budget constraints, and ways to enable safety and security in an AI educational environment. The aim is to inspire and excite educators and school administrators about the potential of AI in education, and to make them feel valued and integral to this transformative journey.

11.1 Infrastructure and Resources for AI Adoption in Schools

The successful integration of AI in the classroom depends on the availability of proper infrastructure and resources. This includes hardware, software, and, most importantly, professional development and support for educators, who play a crucial role in this transformation.

Assessing Infrastructure Needs

A detailed assessment of the existing infrastructure is paramount prior to the introduction of AI tools. This step is crucial as it paves the way for a seamless transition, minimizes disruptions, and identifies and addresses the necessary upgrades or additions.

Example: Infrastructure Assessment in a Middle School

Context: A middle school plans to implement AI-driven personalized learning tools.

Implementation:

Network Evaluation: The school's IT team evaluates the current network infrastructure to ensure it can support AI tools' increased bandwidth and connectivity needs.

Hardware Inventory: An inventory of existing hardware (computers, tablets, smartboards) is conducted to determine if upgrades or additional devices are needed.

Software Compatibility: The school assesses the compatibility of existing software with new AI applications and identifies any necessary upgrades or new software acquisitions.

Outcome: The assessment reveals that the school needs to upgrade its Wi-Fi network, acquire additional tablets for students, and purchase licenses for specific AI software. This proactive approach ensures that the necessary infrastructure is in place before the AI tools are implemented.

Resource Allocation

Effective resource allocation is crucial for supporting AI adoption. This includes funding for technology, professional development, and ongoing maintenance.

Example: Resource Allocation in a High School

Context: A high school plans to integrate AI-powered learning analytics to track student progress and personalize learning.

Implementation:

Budget Planning: The school administration allocates a portion of the budget for purchasing AI tools and upgrading infrastructure.

Grant Applications: The school applies for grants from educational foundations and technology companies to secure funding.

Professional Development: Funds are allocated for professional development workshops and training sessions to help teachers learn how to use the new AI tools effectively.

Outcome: The high school successfully secures grants and allocates budget resources to purchase AI tools, upgrade infrastructure, and provide professional development. This comprehensive approach ensures that teachers are well-prepared and that the necessary resources are available for successful AI integration.

Professional Development

Emphasizing the significance of professional development for educators in AI integration is crucial. It equips them with the necessary skills to effectively use AI tools and integrate them into their teaching practices.

Example: Professional Development Workshops

Context: A school district plans to implement AI-driven personalized learning platforms across its schools.

Implementation:

Workshops and Training: The district organizes workshops and training sessions for teachers, focusing on how to use AI tools to personalize learning and track student progress.

Ongoing Support: The district provides ongoing support through online resources, help desks, and peer mentoring programs.

Certification Programs: Teachers are encouraged to participate in certification programs that provide in-depth training on AI applications in education.

Outcome: Teachers across the district gain the skills and confidence needed to integrate AI tools into their classrooms. The professional development program ensures educators can leverage AI technology to enhance student learning.

Community and Parental Involvement

Underscoring the active involvement and engagement of the community and parents in AI integration is key. It fosters a supportive environment, builds trust, and ensures that AI initiatives align with the community's needs and expectations.

Example: Community and Parental Involvement in AI Integration

Context: An elementary school plans to introduce AI-powered reading interventions for struggling students.

Implementation:

Information Sessions: The school organizes information sessions for parents and community members to explain the benefits of AI-powered reading interventions and how they will be implemented.

Feedback Mechanisms: The school establishes feedback mechanisms by administrating surveys and conducting focus groups to gather input and address concerns from parents and the community.

Regular Updates: The school provides regular updates on the progress and impact of the AI interventions through newsletters, social media, and parent-teacher meetings.

Outcome: Parents and community members are well-informed and supportive of the AI-powered reading interventions. Their input and feedback help refine the implementation process, ensuring that the interventions meet the needs of students and families.

11.2 Overcoming Budget Constraints: Finding and Implementing Free AI Tools

Budgetary limitations can be a challenge, but they should not be a barrier to implementing AI in schools. We want to reassure the audience that a wealth of free or low-cost AI tools is available, and numerous grants and funding opportunities exist. This reassurance is aimed at instilling a sense of hope and confidence, and to make the audience feel that AI integration is within reach, regardless of budget constraints.

Free and Low-Cost AI Tools

Accentuating the potential of identifying and leveraging free or low-cost AI tools in schools is essential. It can help overcome budget constraints while providing valuable AI-driven learning experiences.

Example: Free AI Tools for Education

Tool: Google AI Experiments

Google AI Experiments offers a range of free AI-powered experiments that can be used in the classroom to teach students about AI and machine learning.

Implementation:

Interactive Lessons: Teachers create interactive lessons to engage students in hands-on learning about AI concepts through AI experiments.

Project-Based Learning: Students work on projects using AI experiments, such as creating art with neural networks or training models to recognize images.

Cross-Curricular Integration: The AI experiments are integrated into various subjects, including science, art, and computer science.

Case Study: Google AI Experiments in a Middle School

Context: A middle school uses Google AI Experiments to introduce students to AI and machine learning concepts.

Implementation: Teachers incorporate AI experiments into their lesson plans, allowing students to explore and interact with AI tools.

Outcome: Students develop a basic understanding of AI and machine learning through engaging, hands-on activities. The free tools provide valuable learning experiences without impacting the school's budget.

Grants and Funding

Applying for grants and seeking funding opportunities can provide additional resources for implementing AI in schools.

Example: Applying for Educational Grants

Context: A high school seeks funding to implement AI-driven personalized learning platforms.

Implementation:
  Grant Research: The school's administration researches available grants from educational foundations, government agencies, and technology

companies.

Grant Applications: The school submits detailed grant applications outlining the benefits of AI-driven personalized learning and the school's specific needs.

Community Partnerships: The school secures additional funding by establishing partnerships with local business groups and community organizations.

Case Study: Grant Funding for AI Integration

Context: A high school receives a grant from an educational foundation to implement AI-driven personalized learning platforms.

Implementation: The school uses the grant funds to purchase AI tools, upgrade infrastructure, and provide professional development for teachers.

Outcome: The grant funding enables the school to implement AI-driven personalized learning, improving student engagement and learning outcomes. The program's success attracts additional funding and support from the community.

Creative Budgeting Strategies

Creative budgeting strategies can help schools allocate resources effectively for AI implementation.

Example: Resource Reallocation

Context: An elementary school plans to implement AI-powered reading interventions but faces budget constraints.

Implementation:

Resource Reallocation: The school reallocates existing resources, such as technology and professional development funds, to prioritize AI implementation.

Phased Implementation: The school adopts a phased implementation strategy, starting with a pilot program and gradually expanding to all classrooms.

Partnerships: The school partners with local universities and tech companies to access additional resources and expertise.

Case Study: Phased Implementation of AI-Powered Reading Interventions

Context: An elementary school implements AI-powered reading interventions using a phased approach.

Implementation: The school reallocates resources to fund a pilot program and gradually expands the interventions to all classrooms.

Outcome: The phased implementation allows the school to manage budget constraints while providing valuable AI-driven reading interventions. The pilot program's success attracts additional support and funding, enabling further expansion.

11.3 Creating a Culture of Innovation: Encouraging AI Exploration Among Students

Nurturing a spirit of innovation is essential for students to meaningfully explore AI and develop skills that will prepare them for the future.

Fostering Curiosity

Innovative thinking is nurtured through an environment that fosters curiosity and encourages exploration.

Example: AI Exploration Days

Context: A high school organizes AI Exploration Days to spark students' interest in AI and technology.

Implementation:

Interactive Workshops: The school organizes interactive workshops where students can learn about AI concepts and experiment with AI tools.

Guest Speakers: Experts in AI and technology are invited to speak to students about their work and the potential of AI.

Hands-On Projects: Students on projects that provide hands-on learning and involve developing and applying AI models.

Case Study: AI Exploration Days in a High School

Context: A high school organizes AI Exploration Days to introduce students to AI and encourage exploration.

Implementation: The school hosts workshops, invites guest speakers, and facilitates hands-on projects for students.

Outcome: Students develop a keen interest in AI and technology. The interactive and engaging activities inspire them to pursue further learning and exploration in AI.

Student-Led Projects

Encouraging student-led projects empowers students to take ownership of their learning and explore AI applications.

Example: AI Robotics Projects

Context: A middle school STEM club encourages students to work on AI robotics projects.

Implementation:

Robotics Kits: The school provides robotics kits with sensors and programmable components.

AI Programming: Students learn to program AI models using tools like TensorFlow or Python.

Project Showcase: Students showcase their AI-driven robots in a school-wide exhibition, sharing their projects with peers and the community.

Case Study: AI Robotics Projects in a Middle School STEM Club

Context: A middle school STEM club engages students in AI robotics projects.

Implementation: Students use robotics kits and programming tools to develop AI-driven robots.

Outcome: Students gain practical experience with AI and robotics, developing critical thinking and problem-solving skills. The project showcase provides a platform for students to share their work and inspire others.

Collaboration with Tech Companies

Partnerships with technology companies can provide valuable resources, expertise, and opportunities for students to explore AI.

Example: Tech Company Partnerships

Context: A high school partners with a local tech company to provide students access to AI tools and mentorship.

Implementation:

Access to Tools: The tech company provides students with AI tools and software access.

Mentorship Programs: Employees from the tech company mentor students, guiding them through AI projects and sharing their expertise.

Internship Opportunities: The partnership includes internship opportunities for students to gain hands-on experience in the tech industry.

Case Study: Tech Company Partnership in a High School

Context: A high school partners with a local tech company to enhance AI education.

Implementation: Students can access AI tools and mentorship from tech company employees. Internship opportunities provide real-world experience.

Outcome: Students develop a deeper understanding of AI and its applications. The partnership provides valuable resources and expertise, enhancing the educational experience. Students gain practical skills and insights into potential career paths in technology.

11.4 Safety and Security: Protecting Students in an AI-Enabled Educational Environment

Ensuring students' safety and security is paramount when integrating AI into the classroom. This includes addressing data privacy concerns, implementing best practices for safety, and promoting digital citizenship.

Understanding Risks

AI tools can pose data privacy, security, and ethical risks. Understanding these risks and mitigating them via proactive measures is essential.

Example: Data Privacy Concerns

Context: A school plans to implement AI-powered learning analytics to track student progress.

Implementation:

Data Privacy Assessment: The school conducts a data privacy assessment to identify potential risks and vulnerabilities.

Privacy Policies: The school develops and implements robust data privacy policies to protect student information.

Parental Consent: Parents are informed about using AI tools and data collection, and their consent is obtained.

Case Study: Data Privacy Assessment in a School

Context: A school conducts a data privacy assessment before implementing AI-powered learning analytics.

Implementation: The school identifies potential risks and vulnerabilities, develops privacy policies, and obtains parental consent.

Outcome: The assessment ensures that student data is protected, and implementing privacy policies builds trust with parents and the community. The proactive approach mitigates potential risks and provides a secure learning environment.

Best Practices for Safety

Implementing best practices for safety helps maintain a secure and supportive learning environment when using AI tools.

Example: Safe Use of AI Tools

Context: A school district implements AI tools to enhance personalized learning.

Implementation:

Training and Guidelines: Teachers receive training on the safe use of AI tools and are provided with guidelines.

Monitoring and Oversight: The school district establishes monitoring and oversight mechanisms to use AI tools responsibly.

Incident Response Plan: An incident response plan addresses safety or security issues that arise.

Case Study: Safe Use of AI Tools in a School District

Context: A school district implements AI tools focusing on safety and security.

Implementation: Teachers receive training, establish guidelines, and implement monitoring mechanisms. An incident response plan is developed.

Outcome: Implementing best practices ensures a safe and secure learning environment. Teachers feel confident using AI tools responsibly, and any issues that arise are promptly addressed.

Data Protection

Protecting student data is critical when using AI tools. This includes implementing measures to ensure data security and compliance with relevant laws and regulations.

Example: Data Protection Measures

Context: A school uses AI-powered personalized learning platforms that collect student data.

Implementation:

Encryption: Student data is encrypted to protect it from unauthorized access.

Access Controls: Access to student data is restricted to authorized personnel only.

Compliance: The school ensures compliance with relevant data protection laws and regulations, such as FERPA and GDPR.

Case Study: Data Protection in a School

Context: A school implements data protection measures for AI-powered personalized learning platforms.

Implementation: Student data is encrypted, access controls are established, and data protection laws are complied with.

Outcome: The data protection measures safeguard student information and ensure the data remains private, secure, and protected from unauthorized access. The school's commitment to data protection builds trust with parents and the community.

Digital Citizenship

Promoting digital citizenship is essential for teaching students how to use AI tools responsibly and ethically.

Example: Digital Citizenship Programs

Context: A middle school implements a digital citizenship program to teach students about the responsible use of AI and technology.

Implementation:

Curriculum Integration: Digital citizenship concepts are integrated into the curriculum, focusing on responsible and ethical use of AI and technology.

Workshops and Activities: The school organizes workshops and activities to engage students in discussions about digital citizenship.

Role-Playing Scenarios: Students participate in role-playing scenarios to explore ethical dilemmas and practice making responsible decisions.

Case Study: Digital Citizenship Program in a Middle School

Context: A middle school implements a digital citizenship program to promote responsible use of AI and technology.

Implementation: Digital citizenship concepts are integrated into the curriculum, and workshops and activities are organized.

Outcome: Students develop a strong understanding of the responsible and ethical use of AI and technology. The program fosters a culture of digital citizenship, preparing students to use AI tools safely and responsibly.

Conclusion

Building an AI-ready classroom requires careful planning, strategic implementation, and the right resources. Schools can successfully integrate AI tools into their educational practices by assessing infrastructure needs, allocating resources effectively, providing professional development, and engaging the community. Overcoming budget constraints through free tools, grants, and creative budgeting strategies ensures that all schools can benefit from AI technology regardless of financial limitations. Fostering a culture of innovation encourages students to explore AI and develop skills for the future. Ensuring safety and security through data protection, best practices and digital citizenship programs creates a secure and supportive learning environment. This chapter's detailed examples and case studies illustrate how schools can successfully build AI-ready classrooms, enhancing teaching and learning outcomes for all students.

14

Chapter 12: Case Studies and Real-World Examples

E xploring AI's transformative power in educational settings can ignite inspiration and hope for further adoption. This chapter unveils detailed case studies and real-world examples illuminating AI's diverse applications in education, from empowering struggling students to amplifying teacher effectiveness and fostering global learning communities.

12.1 Success Story: AI in a Rural Classroom

Integrating AI into rural classrooms presents unique challenges, including limited access to technology and connectivity issues. However, with strategic planning and community support, AI can significantly enhance educational outcomes in these settings.

Overcoming Connectivity Issues

Example: Offline AI Tools

Tool: PocketLab

PocketLab is a platform that offers offline AI tools designed for use in environments with limited internet connectivity. It allows students to collect and analyze scientific data using sensors and AI-driven analysis tools without a continuous internet connection.

Implementation:
  - **Equipment Distribution:** PocketLab sensors and devices are distributed to students in a rural school.
  - **Offline Data Collection:** Students use the sensors to conduct scientific experiments and collect data in the field.
  - **AI Analysis:** Data is analyzed using AI algorithms embedded in the PocketLab devices, providing insights and feedback even without internet access.

Case Study: PocketLab in a Rural School

Context: A rural school with limited internet connectivity implemented PocketLab to support science education.

Implementation: Students conducted experiments using PocketLab sensors, collected data, and analyzed it using the AI tools provided by the platform.

Outcome: Students gained hands-on experience with scientific inquiry and data analysis. PocketLab's offline capabilities ensured that learning continued despite connectivity challenges. Teachers reported increased student engagement and understanding of scientific concepts.

Impact on Learning

Example: AI-Powered Reading Programs

Tool: Amira Learning

Amira Learning is an AI-powered reading assistant that helps students enhance their reading skills and comprehension. It effectively supports struggling readers by providing personalized feedback and practice.

Implementation:
- **Initial Assessment:** Students complete an initial reading assessment using the Amira Learning platform.
- **Personalized Practice:** The AI assigns personalized reading exercises based on the assessment results, targeting specific areas for improvement.
- **Progress Monitoring:** Teachers monitor student progress through the platform's analytics dashboard and adjust instruction as needed.

Case Study: Amira Learning in a Rural Elementary School

Context: A rural elementary school implemented Amira Learning to support struggling readers.

Implementation: Students used Amira Learning for daily reading practice, receiving personalized feedback and support from the AI.

Outcome: Students showed significant improvement in reading fluency and comprehension. Teachers reported that the personalized practice provided by Amira Learning effectively addressed individual learning needs, leading to better overall reading outcomes.

Challenges and Solutions

Example: Community Engagement and Support

Tool: Parent and Community Workshops

Schools can engage parents and the community through workshops and information sessions to address challenges such as limited resources and resistance to new technology.

Implementation:
   - **Workshops:** The school organizes workshops to demonstrate the benefits of AI tools and how they will be used to support student learning.
   - **Feedback Sessions:** Parents and community members are invited to provide feedback and ask questions about the AI implementation.
   - **Ongoing Communication:** The school regularly communicates with parents and the community through newsletters, social media, and meetings.

Case Study: Community Engagement in a Rural School

Context: A rural school faced resistance to implementing AI tools due to technology and resource allocation concerns.

Implementation: The school fostered a collaborative environment by organizing workshops and feedback sessions, inviting parents and the community to share their concerns and understand the benefits of AI in education. This inclusive approach led to a smoother implementation and greater acceptance of AI tools.

Outcome: The active engagement of parents and the community through workshops and feedback sessions played a crucial role in successfully integrating AI tools in a rural school. The workshops helped alleviate concerns and build trust, leading to smoother implementation and greater acceptance of AI tools.

12.2 Overcoming Challenges: A Case Study of AI for Special Needs Education

AI has significant potential to support students with special needs by providing personalized learning experiences and assisting with communication and behavioral challenges.

Tailored AI Solutions

Example: AI-Powered Speech Therapy

Tool: Speech Blubs

Speech Blubs is an AI-powered speech therapy app for children with speech and language delays. It provides interactive exercises and personalized feedback to help improve speech skills.

Implementation:
- **Initial Assessment:** The app assesses the child's speech and language abilities to create a personalized therapy plan.
- **Interactive Exercises:** Children engage in fun and interactive exercises designed to improve specific speech and language skills.
- **Progress Tracking:** Parents and therapists can track progress through

the app's analytics dashboard and adjust the therapy plan as needed.

Case Study: Speech Blubs in Special Education

Context: A special education program implemented Speech Blubs to support students with speech and language delays.

Implementation: Students used the app for daily speech therapy sessions, guided by their therapists and supported by their parents.

Outcome: The implementation of Speech Blubs in special education led to significant improvements in speech and language skills. The app's personalized and interactive nature made therapy sessions more engaging and effective. This positive outcome encouraged therapists and parents, fostering increased collaboration and better tracking of progress.

Collaboration with Specialists

Example: AI-Assisted Behavioral Analysis

Tool: Brain Power

Brain Power is an AI-powered platform that provides behavioral support for students with autism. It analyzes behavioral patterns to provide insights and recommendations for interventions.

Implementation:
 - **Behavioral Analysis:** The AI analyzes student behavior using data from wearable devices and classroom observations.
 - **Personalized Interventions:** The platform recommends personalized interventions based on behavioral analysis.

- **Collaboration:** Teachers, therapists, and parents collaborate to implement and adjust interventions as needed.

Case Study: Brain Power in Special Education

Context: A special education program implemented Brain Power to support students with autism.

Implementation: Students used wearable devices to collect data on their behavior. The AI analyzed the data and provided recommendations for personalized interventions.

Outcome: The implementation of Brain Power in a special education program led to significant improvements in social and behavioral skills among students with autism. The AI-assisted analysis helped teachers and therapists develop more effective interventions, leading to improved student outcomes. The shared use of the platform enhanced collaboration between teachers, therapists, and parents.

Measurable Improvements

Example: AI-Powered Learning Tools for Dyslexia

Tool: Bookshare

Bookshare is an AI-powered platform that provides accessible eBooks for students with dyslexia and other reading disabilities. It offers customizable reading experiences to support different learning needs.

Implementation:
- **Accessible eBooks:** Students access eBooks with customizable text and

audio options to accommodate their reading preferences.

- **Reading Support:** The AI provides tools such as text-to-speech, synchronized highlighting, and dictionary support to assist with reading comprehension.

- **Progress Monitoring:** Teachers and parents monitor student progress through the platform's analytics dashboard.

Case Study: Bookshare in Special Education

Context: A special education program implemented Bookshare to support students with dyslexia.

Implementation: Students used Bookshare to access their textbooks and reading materials, utilizing the AI-powered tools for support.

Outcome: Students demonstrated increased reading comprehension and engagement. The customizable reading options and support tools made reading more accessible and enjoyable. Teachers and parents reported that Bookshare significantly improved students' confidence and academic performance.

12.3 Innovative Teaching: AI-Powered Projects in High School

AI-powered projects in high school settings can foster innovation, creativity, and critical thinking among students. These projects provide hands-on experience with AI technologies and prepare students for future careers.

Project-Based Learning

Example: AI-Driven Environmental Science Projects

Tool: IBM Watson

IBM Watson provides AI-powered data analysis and visualization tools, which can be used in environmental science projects to analyze and interpret complex data sets.

Implementation:
- **Data Collection:** Students collect environmental data, such as air quality, water samples, and weather patterns.

AI Analysis: Students analyze the data using IBM Watson to identify patterns, trends, and potential environmental impacts.

- **Presentation:** Students present their findings using visualizations and reports generated by the AI tools.

Case Study: IBM Watson in a High School Environmental Science Class

Context: A high school environmental science class used IBM Watson to analyze ecological data as part of a project-based learning initiative.

Implementation: Students collected data from their local environment and used IBM Watson to analyze and interpret it.

Outcome: Students gained hands-on experience with data analysis and AI technologies. The project fostered critical thinking and problem-solving skills as students developed and presented their findings. Using AI tools enhanced their understanding of environmental science and the importance of data-driven decision-making.

Student Engagement

Example: AI-Powered Creative Writing Projects

Tool: Grammarly

Grammarly is an AI-powered writing assistant providing real-time feedback for grammar, style, and clarity. It can support creative writing projects in high school English classes.

Implementation:
   - **Writing Assignments:** Students complete creative writing assignments using Grammarly for real-time feedback and suggestions.
   - **Peer Review:** Students use Grammarly to review and provide feedback on their peers' writing.
   - **Revisions:** Based on the feedback from Grammarly and peers, students revise and improve their writing.

Case Study: Grammarly in a High School English Class

Context: A high school English class used Grammarly to support creative writing assignments and peer review.

Implementation: Students wrote stories and essays, using Grammarly for real-time feedback and peer review.

Outcome: Students demonstrated improved writing skills and increased engagement with the creative writing process. Grammarly's real-time feedback helped students identify and correct errors, leading to higher-quality writing. Peer review activities fostered collaboration and critical thinking.

Preparing for the Future

Example: AI-Powered STEM Projects

Tool: TensorFlow

TensorFlow is an open-source AI library for developing and deploying machine learning models. It is beneficial for STEM projects in high school.

Implementation:

- **Model Development:** Students learn to develop machine learning models using TensorFlow, focusing on real-world applications such as image recognition or predictive analytics.
- **Project-Based Learning:** Students work on STEM projects that collect data, train models, and interpret results.
- **Showcase:** Students showcase their AI-powered projects in a school-wide exhibition, demonstrating their understanding and skills.

Case Study: TensorFlow in a High School STEM Club

Context: A high school STEM club used TensorFlow to develop machine learning models for various projects.

Implementation: Students learned to use TensorFlow to create models for image recognition and predictive analytics. They worked on projects such as predicting stock market trends and recognizing objects in images.

Outcome: Students gained hands-on experience with machine learning and AI technologies. The projects fostered innovation and critical thinking, preparing students for future careers in STEM fields. The showcase event highlighted their achievements and inspired other students to explore AI.

12.4 Global Perspectives: How Different Countries are Embracing AI in Education

Understanding how countries adopt and integrate AI in education can provide valuable insights and best practices. This section explores case studies from various countries, highlighting their approaches and successes.

Comparative Analysis

Example: AI Integration in Finland

Tool: Elements of AI

Elements of AI is a free online course developed by the University of Helsinki and Reaktor. It is designed to introduce the basics of AI to students and the general public.

Implementation:
  - **Course Enrollment:** Students across Finland enroll in the Elements of AI course, which covers fundamental AI concepts and applications.
  - **Blended Learning:** Schools integrate online courses with classroom instruction, allowing students to apply their learning in practical projects.
  - **Community Engagement:** The course encourages community engagement and collaboration, with students discussing AI concepts and sharing their projects.

Case Study: Elements of AI in Finnish Schools

Context: Finnish schools adopted the Elements of AI course to introduce students to AI and its applications.

Implementation: Students enrolled in the online course and participated in blended learning activities, combining online learning with classroom projects.

Outcome: Students developed a strong foundation in AI concepts and applications. The course fostered curiosity and engagement with AI, preparing students for future studies and careers in technology. The success of the program inspired other countries to adopt similar approaches.

Success Stories from Around the World

Example: AI-Powered Education in China

Tool: Squirrel AI Learning

Squirrel AI Learning is an AI-powered adaptive learning platform widely used in China to personalize education and improve student outcomes.

Implementation:
  - **Personalized Learning:** The AI platform provides personalized learning paths based on student performance and learning styles.
  - **Real-Time Feedback:** Students receive real-time feedback and support from the AI, helping them address gaps in their understanding.
  - **Data-Driven Insights:** Teachers use the platform's data-driven insights to tailor instruction and support individual students.

Case Study: Squirrel AI Learning in Chinese Schools

Context: Chinese schools implemented Squirrel AI Learning to personalize education and improve student outcomes.

Implementation: Students used the AI platform for personalized learning, receiving real-time feedback and support. Teachers used data-driven insights to tailor their instruction.

Outcome: Students demonstrated significant improvement in academic performance and engagement. The AI platform's personalized learning paths and real-time feedback helped effectively address individual learning needs. The success of Squirrel AI Learning has led to its adoption in other countries.

Global Challenges

Example: Addressing the Digital Divide

Tool: UNESCO's Global Education Coalition

UNESCO's Global Education Coalition is an initiative that bridges the digital divide and ensures inclusive education through collaboration among governments, technology companies, and educational organizations.

Implementation:
- **Partnerships**: The coalition partners with governments and organizations to enable internet and technology access to underserved regions.
- **Digital Literacy Programs:** The coalition offers digital literacy programs to help students and teachers use technology effectively.
- **Resource Sharing:** The coalition facilitates sharing educational resources and best practices among member countries.

Case Study: UNESCO's Global Education Coalition

Context: UNESCO's Global Education Coalition addresses the digital divide by providing technology and digital literacy programs to underserved regions.

Implementation: The coalition partners with governments and organizations to provide technology, internet access, and digital literacy programs.

Outcome: Increased access to technology and digital literacy in underserved regions. The coalition's efforts have helped bridge the digital divide, ensuring all students have access to quality education. Sharing resources and best practices has fostered international collaboration and improved educational outcomes globally.

International Collaboration

Example: AI Research and Development Partnerships

Tool: AI4Edu

AI4Edu is a collaborative initiative that brings together researchers, educators, and technology companies to develop and implement AI solutions for education.

Implementation:
 - **Research Collaborations:** The initiative facilitates research collaborations to develop AI tools and solutions for education.
 - **Pilot Programs:** AI4Edu supports pilot programs to test and refine AI tools in real-world educational settings.
 - **Knowledge Sharing:** The initiative promotes knowledge sharing and disseminating research findings and best practices.

Case Study: AI4Edu Research Collaboration

Context: AI4Edu facilitates international research collaborations to develop AI tools for education.

Implementation: Researchers, educators, and technology companies collaborate on developing and testing AI solutions. Pilot programs are implemented in schools to refine the tools.

Outcome: The collaboration has led to developing innovative AI tools and solutions that enhance teaching and learning. The pilot programs provide valuable insights into the practical implementation of AI in education. The knowledge-sharing and dissemination efforts have helped spread best practices globally.

Conclusion

The case studies and real-world examples presented in this chapter illustrate the diverse ways AI is being integrated into education across different contexts. From supporting struggling students in rural classrooms to enhancing special needs education and fostering innovation in high school projects, AI has the potential to transform education. Understanding how different countries and educational settings successfully implement AI can provide valuable insights and inspiration for further adoption. As AI technology evolves, educators, administrators, and policymakers must stay informed, collaborate, and adapt to ensure technology's ethical and responsible use to improve educational outcomes for all students.

15

Conclusion and Way Forward

As we conclude this journey through the integration of AI in education, it is evident that AI holds transformative potential for enhancing teaching and learning. Throughout this book, we have explored various facets of AI in education, from understanding its fundamental concepts to addressing ethical considerations and practical implementation strategies. Here, we summarize the key insights and provide a vision for the future.

Key Insights

1. Demystifying AI for Educators: AI can seem complex and intimidating, but breaking its concepts and applications into understandable terms makes it accessible. Educators must feel confident in understanding AI to leverage its full potential.

2. Ethical Considerations: Ethical use of AI in education is paramount. Transparency, fairness, and accountability are critical to building trust and ensuring that AI applications benefit all students equitably.

3. AI Integration Strategies: Effective integration of AI requires thoughtful planning, including infrastructure assessments, resource allocation, and professional development. Schools must prepare both technologically and culturally for AI adoption.

4. Personalized Learning: AI-driven personalized learning tailors educational experiences to individual student needs, improving engagement and outcomes. Examples like Knewton Alta and DreamBox Learning illustrate the potential of personalized learning on student empowerment.

5. Teacher Support: AI can enhance teacher effectiveness by automating routine tasks, providing real-time feedback, and offering data-driven insights. Tools like Gradescope and Amira Learning exemplify how AI supports teachers in delivering high-quality education.

6. Global Classroom: AI facilitates global learning communities, breaking geographical barriers and promoting cultural exchange. Initiatives like ePals and Google Classroom demonstrate the potential for international collaboration in education.

7. Special Needs Education: AI provides valuable support for students with special needs, offering personalized interventions and enhancing communication. Tools like Speech Blubs and Brain Power show how AI can significantly impact special education.

8. Innovation and Creativity: Encouraging students to explore AI through project-based learning fosters innovation and critical thinking. Examples from high school STEM clubs and AI-driven projects highlight the importance of hands-on learning.

9. Professional Development: Ongoing professional development is essential for educators to stay updated on AI advancements and integrate AI successfully into their teaching practices. Programs like those offered by edX and ISTE support continuous learning.

Way Forward

As we look to the future, it is crucial to build on the foundations laid out in this book and continue to advance the integration of AI in education. Here are some actionable steps and strategies for educators, administrators, and policymakers:

1. Commit to Continuous Learning: AI technology is rapidly evolving. Continuous learning and professional development in AI advancements and best practices among educators is critical.

2. Foster a Culture of Innovation: Schools should create an environment encouraging experimentation and innovation. This includes supporting student-led AI projects, organizing AI exploration days, and fostering collaboration with technology companies.

3. Prioritize Equity and Access: Ensure that AI learning environments are accessible to all students, regardless of their socio-economic background. This involves addressing the digital divide and providing necessary infrastructure and support in underserved communities.

4. Implement Robust Data Privacy Measures: Protecting student data is paramount. Schools must implement robust data privacy measures, comply with relevant regulations, and communicate transparently with parents and the community about data usage.

5. Engage Stakeholders: Actively engage parents, community members, and other stakeholders in the AI integration process. Regular communication, feedback mechanisms, and community involvement are essential for building trust and support.

6. Develop Ethical Guidelines: Establish and adhere to ethical guidelines for using AI in education. This includes ensuring transparency, minimizing bias, and maintaining accountability in AI applications.

7. Leverage Free and Low-Cost AI Tools: Explore and implement free or low-cost AI tools to overcome budget constraints. Platforms like Google AI Experiments and Khan Academy offer valuable resources without significant financial investment.

8. Collaborate Globally: Embrace international collaboration to share best practices, resources, and insights on AI in education. Initiatives like UNESCO's Global Education Coalition can provide valuable support and foster global learning communities.

9. Prepare Students for the Future Empower students with the skills and knowledge needed to thrive in an AI-driven world. This includes teaching AI literacy, promoting critical thinking, and encouraging exploration and innovation.

Vision for the Future

The future of education is bright with the integration of AI. As we move forward, it is essential to embrace the potential of AI to enhance learning, support teachers, and create more inclusive and engaging educational environments. By prioritizing equity, ethics, and continuous learning, we can ensure that AI is a powerful tool for improving education for all students.

Let us envision a future where:

Personalized Learning: Students receive a personalized learning experience customized to address their unique learning needs and styles, which inspires them and enhances their engagement and outcomes.

- **Teacher Empowerment:** AI tools empower teachers by streamlining administrative tasks, providing real-time insights, and supporting differentiated instruction.

- **Global Collaboration:** Students and educators collaborate across borders, sharing knowledge and cultural experiences and creating a global community of learners.

- **Inclusive Education:** AI supports diverse learners, including those with special needs, by providing tailored interventions and accessible learning resources.

- **Ethical AI Use:** AI applications in education are developed and used ethically, ensuring fairness, transparency, and accountability.

- **Innovative Learning:** Classrooms are dynamic and innovative, with students actively exploring and creating with AI, preparing them for future careers and challenges.

By working together, educators, administrators, policymakers, and the community can unleash the potential of AI to transform education. Let us commit to building an AI-ready education system that prepares our students for a future full of possibilities and opportunities.

Final Thoughts

Integrating AI in education is a technological advancement and a paradigm shift that requires a collaborative and thoughtful approach. As we embark on this journey, remember that the ultimate goal is to enhance learning and provide every student with the best possible education. By embracing AI responsibly and ethically, we can create a brighter future for education and unlock the full potential of our students.

Thank you for joining us on this journey to explore AI in education. Together, we can make a meaningful impact and shape the future of learning for generations.

16

Appendix A: Glossary of AI Terms in Education

This glossary defines key terms related to AI in education, helping readers understand the language and concepts discussed throughout the book.

- **Adaptive Learning:** A method of delivering personalized learning experiences that adjust to the learner's needs based on their performance in real-time.

- **AI-Powered Tools:** Software applications that leverage AI technologies to enhance teaching and learning, such as personalized learning platforms, grading assistants, and virtual tutors.

- **Artificial Intelligence (AI):** The simulation of human intelligence processes by machines, especially computer systems. These processes include learning, reasoning, and self-correction.

- **Data Privacy:** Protect personal data from unauthorized access and use. In education, this includes ensuring the confidentiality and integrity of student data.

- **Ethical AI:** The practice of designing and using AI in ways that are transparent, accountable, and aligned with ethical standards, ensuring fairness and equity.

- **Machine Learning (ML):** A subset of AI that uses algorithms and statistical models to enable computers to improve their performance on tasks through experience.

- **Natural Language Processing (NLP):** A field of AI focusing on the interaction between computers and humans through natural language. It involves the ability of a computer to understand, interpret, and generate human language.

- **Neural Networks:** A series of algorithms that mimic the operations of a human brain to recognize relationships between vast amounts of data.

- **Personalized Learning:** Educational programs that are tailored to the individual needs, skills, and interests of each student, often using AI to adapt the learning experience in real-time.

- **Predictive Analytics:** Using data, statistical algorithms, and machine learning techniques to identify the likelihood of future outcomes based on historical data.

17

Appendix B: AI Tools and Resources for Educators

This appendix lists various AI tools and resources mentioned in the book, along with a brief description and their application in education.

AI Tools

1. **Amira Learning:** An AI-powered reading assistant that helps students improve reading fluency and comprehension.

- **Website**: https://www.amiralearning.com

2. **Bookshare:** An AI-powered platform offering accessible eBooks for students with reading challenges.

- **Website**: https://www.bookshare.org

3. **Brain Power:** An AI platform providing behavioral support for students with autism.

- **Website**: https://www.brain-power.com

4. **Google AI Experiments:** Free AI-powered experiments for hands-on learning about AI concepts.

- **Website**: https://experiments.withgoogle.com/collection/ai

5. **Grammarly**: An AI-powered writing assistant providing real-time feedback for grammar, style, and clarity.

- **Website**: https://www.grammarly.com

6. **IBM Watson for Education:** AI-powered tools for data analysis and visualization in educational projects.

- **Website**: https://www.ibm.com/watson/education

7. **Knewton Alta:** An adaptive learning platform that offers personalized coursework in various subjects.

- **Website**: https://www.knewton.com/alta/

8. **Speech Blubs:** An AI-powered speech therapy app for children with speech and language delays.

- **Website**: https://www.speechblubs.com

9. **Squirrel AI Learning**: An adaptive learning platform widely used in China for personalized education.

- **Website**: https://squirrelai.com/en

10. **TensorFlow:** An open-source AI library to develop and deploy machine

learning models.

- **Website**: https://www.tensorflow.org

Educational Resources

1. **Elements of AI:** A free online course developed by the University of Helsinki and Reaktor to introduce AI basics.

- **Website**: https://www.elementsofai.com

2. **ISTE Standards:** Standards provided by the International Society for Technology in Education for integrating technology in education.

- **Website**: https://www.iste.org/standards

3. **UNESCO's Global Education Coalition:** This initiative bridges the digital divide, ensuring equitable and inclusive access to education.

- **Website**: https://en.unesco.org/covid19/educationresponse/globalcoaliti on

18

Appendix C: Sample Lesson Plans Integrating AI

This appendix provides sample lesson plans that illustrate the integration of AI tools into various subjects and emphasize their practicality and effectiveness in enhancing learning and teaching.

Sample Lesson Plan 1: AI in Environmental Science

Grade Level: High School

Subject: Environmental Science

Objective: To analyze environmental data using AI tools and explore the impact of various environmental factors.

Materials:

- Computers or tablets with internet access

- IBM Watson for Education

- Environmental data sets (e.g., air quality, water samples)

Procedure:

1. **Introduction (10 minutes):** Briefly introduce the concept of AI and its application in environmental science.

2. **Data Collection (30 minutes):** Students collect environmental data from their local environment or use provided data sets.

3. **AI Analysis (40 minutes):** Students use IBM Watson to analyze the collected data, identifying patterns and trends.

4. **Presentation (20 minutes):** Students present their findings using visualizations and reports generated by the AI tools.

5. **Discussion (20 minutes):** Discuss the results and the role of AI in enhancing environmental science research.

Assessment:
- Participation in data collection and analysis
- Quality of AI-generated reports and visualizations
- Effectiveness of the presentation and discussion contributions

Sample Lesson Plan 2: AI-Powered Creative Writing

Grade Level: Middle School

Subject: English Language Arts

Objective: To enhance creative writing skills using AI-powered writing tools.

Materials:
- Computers or tablets with internet access
- Grammarly

Procedure:

1. **Introduction (10 minutes):** Introduce Grammarly and explain how its AI-powered algorithms can assist in the writing process by providing real-time grammar, style, and clarity feedback.

2. **Writing Assignment (30 minutes):** Students write a short story or essay using Grammarly for real-time grammar, style, and clarity feedback.

3. **Peer Review (20 minutes):** Students review and provide feedback on each other's work using Grammarly.

4. **Revisions (20 minutes):** Students revise and improve their writing based on feedback.

5. **Sharing and Discussion (20 minutes):** Students share their revised work and discuss how AI tools helped enhance their writing.

Assessment:
- Quality of the initial and revised writing assignments
- Engagement in peer review activities
- Participation in sharing and discussion

19

Appendix D: Professional Development Resources

This appendix presents a comprehensive list of easily accessible resources, designed to empower educators to confidently pursue professional development in AI and technology integration.

1. **Coursera:** Offers courses on AI and education technology from leading universities and institutions.

- **Website:** https://www.coursera.org

2. **edX Professional Development Programs:** Offers courses on AI and education technology for educators.

- **Website**: https://www.edx.org

3. **Future Learn:** Provides online courses on AI and its applications in education.

- **Website:** https://www.futurelearn.com

4. **ISTE Learning:** Provides online courses, webinars, and workshops on technology integration in education.

- **Website:** https://www.iste.org/learn

5. **TeachAI:** A platform offering resources and training for educators on AI literacy and integration.

- **Website:** https://teachai.org

20

Appendix E: Additional Case Studies and Examples

This appendix includes additional case studies and real-world examples of AI integration in education.

Case Study: AI for Language Learning

Tool: Duolingo

Context: A high school uses Duolingo to support Spanish language learning for students.

Implementation:

- **Daily Practice:** Students complete daily language practice exercises on Duolingo.

- **Classroom Integration:** Teachers integrate Duolingo exercises into the language curriculum.

- **Progress Tracking:** Teachers monitor student progress through Duolingo's analytics dashboard.

Outcome: Students demonstrate improved language proficiency and engagement, with an average increase of [10%] in their Spanish test scores. Duolingo's gamified approach makes learning fun and effective, leading to better retention and performance in language classes.

Case Study: AI for Math Tutoring

Tool: DreamBox Learning

Context: An elementary school implements DreamBox Learning to provide personalized math tutoring.

Implementation:

- **Initial Assessment:** Students complete an initial assessment to determine their math proficiency.

- **Adaptive Learning:** The AI provides personalized math exercises based on the assessment results.

- **Teacher Support:** Teachers use data-driven insights from the platform to support individual students.

Outcome: Implementing DreamBox Learning has led to a dramatic improvement in students' math skills and confidence. The personalized approach has been instrumental in addressing individual learning needs and improving overall math performance. This success story should empower educators to address the unique needs of each student with the help of AI tools.

These case studies illustrate how educators can effectively integrate AI into their classrooms. By leveraging AI tools, educators can enhance their teaching practices and create more interactive and personalized learning experiences for their students. These case studies are intended to reassure educators about the feasibility and effectiveness of integrating AI into their classrooms.

21

References

- Adams Becker, S., Cummins, M., Davis, A., Freeman, A., Hall Giesinger, C., & Ananthanarayanan, V. (2017). NMC/CoSN Horizon Report: 2017 K–12 Edition. Austin, TX: The New Media Consortium.
- Amira Learning. (n.d.). Amira Learning. Retrieved from https://www.amiralearning.com
- Bernard Marr. (2019, October 9). How Is AI Used In Education — Real World Examples Of Today And A Peek Into The Future. Forbes. Retrieved from https://www.forbes.com/sites/bernardmarr/2019/10/09/how-is-ai-used-in-education-real-world-examples-of-today-and-a-peek-into-the-future/?sh=34e12a2b7c2a
- Bookshare. (n.d.). Accessible Books for Individuals with Print Disabilities. Retrieved from https://www.bookshare.org
- Brain Power. (n.d.). Empowering Individuals with Autism and Other Challenges. Retrieved from https://www.brain-power.com
- Clifford, G. (2019). Artificial Intelligence and Education: Learning Algorithms in the Classroom. Journal of Educational Technology, 34(2), 121-135.
- Dillon, G., & Rose, K. (2019). The Impact of AI on Education. Educational Researcher, 48(3), 154-163.

- edX. (n.d.). Online Courses. Retrieved from https://www.edx.org
- Elements of AI. (n.d.). Free Online Course. University of Helsinki and Reaktor. Retrieved from https://www.elementsofai.com
- Future of Privacy Forum. (2018). Artificial Intelligence and Student Privacy: The Role of Data and Privacy Protection in the Age of AI. Retrieved from https://fpf.org/wp content/uploads/2018/10/FPF Artificial-Intelligence-and-Student-Privacy.pdf
- Grammarly. (n.d.). AI-Powered Writing Assistant. Retrieved from https://www.grammarly.com
- Holmes, W., Bialik, M., & Fadel, C. (2019). Artificial Intelligence in Education: Promises and Implications for Teaching and Learning. Boston, MA: Center for Curriculum Redesign.
- IBM Watson Education. (n.d.). IBM Watson for Education. Retrieved from https://www.ibm.com/watson/education
- ISTE. (n.d.). International Society for Technology in Education Standards. Retrieved from https://www.iste.org/standards
- Khan Academy. (n.d.). Free Educational Resources. Retrieved from https://www.khanacademy.org
- Knewton Alta. (n.d.). Adaptive Learning Platform. Retrieved from https://www.knewton.com/alta/
- Luckin, R. (2017). Machine Learning and Human Intelligence: The Future of Education for the 21st Century. London: UCL Institute of Education Press.
- Mariano, R. (2016). From BOOKs to MOOCs: The Integration of Open Educational Resources to Library Services in American Spaces in East Asia and Pacific Region. http://library.ifla.org/1390/
- Marr, B. (2018). Data-Driven Education: How AI and Big Data Are Transforming Education. Retrieved from https://bernardmarr.com/default.asp?contentID=1841
- Microsoft Azure Machine Learning. (n.d.). Machine Learning Service. Retrieved from https://azure.microsoft.com/en-us/services/machine-learning/
- Microsoft Teams for Education. (n.d.). Collaborative Workspace for

Education. Retrieved from https://www.microsoft.com/en-us/educati on/products/teams

- Microsoft Translator. (n.d.). AI-Powered Translation Tool. Retrieved from https://www.microsoft.com/en-us/translator/education
- Popenici, S., & Kerr, S. (2017). Exploring the Impact of Artificial Intelligence on Teaching and Learning in Higher Education. Research and Practice in Technology Enhanced Learning, 12(22).
- PocketLab. (n.d.). Offline AI Tools for Scientific Inquiry. Retrieved from https://www.thepocketlab.com
- Squirrel AI Learning. (n.d.). Adaptive Learning Platform. Retrieved from https://squirrelai.com/en
- Speech Blubs. (n.d.). AI-Powered Speech Therapy App. Retrieved from https://www.speechblubs.com
- TensorFlow. (n.d.). Open Source Machine Learning Platform. Retrieved from https://www.tensorflow.org
- Turnitin. (n.d.). Plagiarism Detection and Originality Check. Retrieved from https://www.turnitin.com
- Udacity. (n.d.). Online Learning Platform. Retrieved from https://ww w.udacity.com
- UNESCO. (2020). AI and Education: Guidance for Policy-makers. Retrieved from https://unesdoc.unesco.org/ark:/48223/pf0000376709
- UNESCO's Global Education Coalition. (n.d.). Global Initiative for Education. Retrieved from https://en.unesco.org/covid19/educationre sponse/globalcoalition

22

Bonus 1: Spotlight on AI in Education and Gender Equity

I ntroduction

Artificial Intelligence (AI) holds immense potential to revolutionize education, paving the way for more personalized and compelling learning experiences. However, it also presents challenges, particularly in gender equity. By ensuring that AI in education is a force for gender equity, we can address biases in AI systems, create inclusive learning environments, and empower all students, regardless of gender, to thrive. This bonus section is a beacon of hope, focusing on the intersection of AI in education and gender equity, offering insights and strategies for educators to foster an inclusive educational experience.

Addressing Gender Bias in AI

Understanding Gender Bias in AI

AI systems, while revolutionary, can inadvertently perpetuate gender biases present in their training data. For instance, if an AI tool is trained on data that reflects historical gender biases, it might reinforce those biases in its recommendations and decisions. Recognizing the role of AI in perpetuating biases is a crucial step toward promoting gender equity in education.

Examples of Gender Bias:

- **Voice Recognition:** AI voice recognition systems often perform better for male voices because they are typically trained on male-dominated datasets.

- **Career Recommendations:** AI-driven career recommendation systems might suggest stereotypically gendered career paths, such as engineering for boys and nursing for girls.

Strategies to Mitigate Gender Bias

1. Diverse Training Data:

- Ensure that AI systems are trained on diverse and representative datasets that include balanced gender representation.

- Regularly update datasets to reflect changing societal norms and gender roles.

Example: A school using an AI-driven career guidance tool should include diverse career profiles and paths that challenge traditional gender roles, showcasing female engineers and male nurses to inspire all students.

2. Bias Audits:

- Conduct regular audits of AI systems to identify and mitigate gender biases.

- Involve diverse stakeholders in the auditing process to provide multiple perspectives.

Example: An audit of an AI grading tool at a high school revealed a bias: Female students' essays were scored lower because the tool's language model was trained on biased datasets. The school retrained the AI with a more balanced dataset, resulting in fairer grading outcomes.

3. Inclusive AI Design:

- Design AI tools with inclusivity in mind, considering how different genders might interact with the technology.

- Use gender-neutral language and interfaces.

Example: An AI-driven learning platform for language arts can use gender-neutral avatars and names to ensure all students feel represented and included.

Promoting Gender Equity through AI

Personalized Learning Paths

With its potential to create personalized learning experiences, AI can cater to each student's unique needs and interests, regardless of gender. By tailoring content to individual preferences, AI can encourage students to explore subjects they might not traditionally consider. This potential of AI in creating inclusive learning environments is a powerful tool in empowering all students to succeed, regardless of gender.

Example: At an elementary school, an AI-driven math tutoring system identified female students underperforming due to a lack of interest. By integrating real-world math applications in fashion and architecture, the AI tool increased engagement and performance among female students.

Encouraging STEM Participation

AI can encourage girls to pursue STEM (Science, Technology, Engineering, Mathematics) by providing engaging and supportive learning environments.

Strategies:

- **Role Models:** Use AI to highlight female role models in STEM fields within educational content.

- **Interactive Learning:** Develop AI-driven interactive learning modules that make STEM subjects more engaging and accessible.

Example: An academy focused on technology education for girls uses an AI-driven platform that features female scientists and engineers in interactive science lessons, inspiring girls to consider careers in STEM.

Supporting Teachers

AI tools can support teachers in promoting gender equity by providing insights into student performance and engagement, allowing for targeted interventions.

Example: An AI analytics tool at a high school tracks student engagement in various subjects. It alerted teachers that female students were less engaged in computer science classes. The school responded by introducing a mentorship program pairing female students with women in tech, significantly boosting engagement and performance.

Case Studies

Case Study 1: AI-Driven Mentorship Programs

Scenario: A high school implements an AI-driven mentorship program to connect female students interested in STEM with female professionals in the field.

Implementation:

- The AI system matches students with mentors based on their interests, career goals, and availability.

- Mentors and mentees engage in regular virtual meetings, with AI providing resources and discussion topics tailored to the mentee's interests.

Outcome: The high school program resulted in a 30% increase in female students enrolling in advanced STEM courses and pursuing STEM careers after graduation.

Case Study 2: Inclusive Curriculum Development

Scenario: An elementary school uses AI to develop an inclusive curriculum highlighting contributions from diverse genders in various fields.

Implementation:

The AI tool curates content from various sources, ensuring balanced gender representation in history, science, literature, and the arts.

- Teachers use this content to create lesson plans that challenge traditional gender roles and stereotypes.

Outcome: The elementary school's inclusive curriculum increased student engagement and participation, with notable improvements in girls' performance and interest in STEM subjects.

Building an Inclusive Learning Environment

Engaging Parents and Community

Building gender equity through AI in education requires the support of parents and the community. Engaging stakeholders ensures a holistic approach to fostering an inclusive learning environment.

Strategies:

- **Workshops and Seminars:** Conduct workshops for parents on AI's benefits and ethical use in promoting gender equity.

- **Community Involvement:** Involve community leaders in discussing AI and gender equity in education.

Example: A high school hosts annual seminars for parents and community members to discuss AI's role in education and gender equity, fostering a supportive student environment.

Continuous Professional Development for Educators

Educators are the key to leveraging AI to promote gender equity. Continuous professional development ensures they have the knowledge and skills to use AI effectively and ethically, empowering them to make a significant impact in their classrooms.

Strategies:

- **Training Programs:** Offer regular training programs on AI tools and their applications in promoting gender equity.

- **Collaborative Learning:** Create opportunities for teachers to collaborate and share best practices on using AI to foster an inclusive classroom.

Example: A school district provides annual AI training sessions for teachers, focusing on using AI to support diverse learning needs and promote gender equity.

Conclusion

AI has the potential to significantly advance gender equity in education by personalizing learning, encouraging participation in STEM, and supporting teachers. However, it's crucial to remember that this potential can only be fully realized if we address gender biases in AI systems and promote inclusive practices. As educators, it's our responsibility to create educational systems where all students, regardless of gender, can succeed. This bonus section has provided strategies and examples to help us harness AI to promote gender equity effectively. As we continue integrating AI into education, let's remain committed to delivering on the principles of inclusivity and fairness, essential for achieving accurate gender equity.

References

- Google AI. (n.d.). Teachable Machine. Retrieved from https://teachable machine.withgoogle.com/
- National Science Foundation. (2020). Women, Minorities, and Persons with Disabilities in Science and Engineering. Retrieved from https://n cses.nsf.gov/pubs/nsf19304/digest
- Obermeyer, Z., Powers, B., Vogeli, C., & Mullainathan, S. (2019). Dissecting racial bias in an algorithm used to manage the health of populations. Science, 366(6464), 447-453. Retrieved from https://scie nce.sciencemag.org/content/366/6464/447
- Paradis, K. (2023). More Than Ones and Zeros: Developing an Intersectional Framework for Artificial Intelligence. Journal of Information Ethics, 32(2), 70-83.
- Roberts, T., Maiorca, C., & Chapman, P. (2019). Equitably engaging all

students in STEM. The Elementary STEM Journal, 23(4), 30-33.

- Ulnicane, I. (2024). Intersectionality in Artificial Intelligence: Framing Concerns and Recommendations for Action. https://doi.org/10.17645 /si.7543
- UNESCO. (2020). Artificial Intelligence in Education: Challenges and Opportunities for Sustainable Development. Retrieved from https://en .unesco.org/themes/education-and-artificial-intelligence
- West, S. M., Whittaker, M., & Crawford, K. (2019). Discriminating Systems: Gender, Race, and Power in AI. AI Now Institute. Retrieved from https://ainowinstitute.org/discriminatingsystems.html

23

Bonus 2: Spotlight on AI in Education and Trust

I ntroduction

Building trust among educators, students, and parents is crucial as Artificial Intelligence (AI) becomes more integrated into educational settings. Trust is not just a nice-to-have but a foundational requirement for successfully implementing AI technologies. Without trust, the potential benefits of AI in education may not be fully realized. This bonus section addresses critical issues related to AI in education and offers strategies for building trust, ensuring ethical use, and fostering a positive perception of AI's role in learning.

Addressing Trust Issues in AI

Transparency

Importance of Transparency:

Transparency is vital in building trust in AI systems. Educators, students, and parents need to understand how AI tools work, what data they collect, and how they make decisions.

Strategies for Transparency:

- **Clear Documentation:** Provide comprehensive, easy-to-understand documentation about how AI tools function, including the algorithms used and data collected.

- **User-Friendly Interfaces:** Design AI tools with interfaces that clearly show how decisions are made and allow users to understand the process.

Open Communication: Regularly communicate with all stakeholders, including educators, students, parents, and school administrators, about AI tools' updates, changes, and benefits. Each stakeholder is responsible for ensuring AI's ethical and practical use in education, and open communication is critical to this process.

Example: At an elementary school, teachers use an AI-driven personalized learning platform with a "How It Works" section that explains the algorithms and data usage in simple terms. Regular newsletters update parents on the AI tool's performance and any changes.

Ethical Considerations

Importance of Ethics:

Ethical considerations are paramount in building trust. Addressing data privacy, bias, and fairness ensures that AI tools are used responsibly. However, it's essential to acknowledge that AI has risks. For instance, AI systems are trained on data. When the data is skewed, AI systems can inadvertently perpetuate biases. Educators can take proactive steps to mitigate these challenges by being aware of them.

Strategies for Ethical AI Use:

- **Data Privacy:** Implement robust data protection and compliance with regulations like FERPA and GDPR.

- **Bias Mitigation:** Regularly audit AI systems for bias and make adjustments to ensure fairness.

- **Fairness and Inclusivity:** Design inclusive AI tools catering to diverse student populations.

Example: A high school uses an AI-driven assessment tool that has been rigorously tested for bias. It conducts regular audits and involves a diverse team of educators in evaluating the tool's fairness.

Accountability

Importance of Accountability:

Holding AI systems and their developers accountable builds trust and ensures that AI tools are used ethically and effectively.

Strategies for Accountability:

- **Regular Audits:** Conduct audits of AI systems to ensure they function as designed and adhere to ethical guidelines.

- **Feedback Mechanisms:** Establish clear channels for students, parents, and teachers to provide feedback on AI tools.

- **Transparent Reporting:** Publicly report on AI tools' performance, audits, and feedback.

Example: A middle school publishes an annual report on its AI tools, detailing performance metrics, audit results, and how it has addressed any issues stakeholders raise. When a concern is raised, the school initiates a thorough investigation involving all relevant parties and takes appropriate actions to rectify the problem. This transparent process reassures stakeholders of the school's commitment to their concerns and the responsible use of AI.

Building Trust Through Education

Educating Stakeholders

Importance of Education:

Educating all stakeholders about AI's capabilities, limitations, and benefits is crucial for building trust.

Strategies for Educating Stakeholders:

- **Workshops and Training:** Conduct workshops for teachers, students, and parents to learn about AI tools and how to use them effectively. These workshops can include hands-on training with AI tools, discussions on the benefits and challenges of AI in education, and interactive sessions to address any questions or concerns.

- **Resources and Guides:** Provide accessible resources and guides explaining AI concepts and the specific tools used.

- **Open Forums:** Host forums and Q&A sessions where communities can engage, learn, ask questions, and express concerns.

Example: A high school hosts quarterly AI workshops for parents, students, and teachers. These sessions cover the basics of AI and its use in the classroom and provide hands-on training with AI tools.

Fostering Positive Perceptions of AI

Highlighting Success Stories

Importance of Success Stories:

Sharing success stories helps demonstrate AI's tangible benefits and build positive perceptions. For instance, AI tools can personalize learning, adapt to unique student needs, and generate real-time feedback, significantly enhancing the learning experience.

Strategies for Highlighting Success Stories:

- **Case Studies:** Develop detailed case studies that showcase how AI has positively impacted student learning and engagement.

- **Testimonials:** Collect and share testimonials from teachers, students, and parents who have had positive experiences with AI tools.

- **Media Coverage:** Leverage school newsletters, local media, and social media to highlight successes.

Example: An elementary school's newsletter featured a success story about how an AI tool helped improve reading scores by 15% over a semester. The story included quotes from students and teachers, providing a personal touch.

Demonstrating Continuous Improvement

Importance of Continuous Improvement

Demonstrating a commitment to continuously improving AI tools shows stakeholders that their feedback is valued and that the school is dedicated to providing the best learning experiences.

Strategies for Continuous Improvement

- **Iterative Updates:** Regularly update AI tools based on feedback and performance data.

- **Pilot Programs**: Run pilot programs for new AI tools to gather feedback and improve them before full implementation.

- **Engage with Developers:** Work closely with AI developers to address issues and ensure the tools meet the school's needs.

Example: A high school runs pilot programs for new AI tools in a few classrooms each semester. They gather detailed feedback from teachers and students, which they use to make continuous improvements before rolling out the tools school-wide. This commitment to adaptability ensures that the tools always meet the school's evolving needs.

Overcoming Common Objections

Objection 1: "AI will invade student privacy."

Solution: This objection is often raised due to the perception that AI tools collect and use personal data without consent. However, it's important to note that AI tools can be designed to prioritize data privacy, and strict adherence to legal regulations can further protect student privacy. Emphasize robust data privacy measures and strict adherence to legal frameworks.

Example: A school district provides detailed information about its data privacy policies and how AI tools comply with FERPA and GDPR. It also uses anonymized data whenever possible to further protect student privacy.

Objection 2: "AI will replace teachers."

Solution: Emphasize that AI is a tool to augment, not replace, teachers. It collaborates with educators, handling routine tasks and freeing teachers to focus on personalized instruction and student engagement.

Example: At a high school, teachers use an AI grading assistant that handles multiple-choice questions, allowing them to invest time in sharing detailed feedback on essays and projects. This innovative tool has significantly improved the efficiency and effectiveness of grading.

Objection 3: "AI tools are too expensive."

Solution: Many AI tools offer free versions or educational discounts. Additionally, schools can apply for funding and grant opportunities to cover costs.

Example: An elementary school secured a grant to implement an AI-driven personalized learning platform. They also use free AI tools like Google's Teachable Machine for classroom projects.

Conclusion

Building trust in AI for education involves transparency, ethical considerations, accountability, and continuous education. By addressing these key areas and overcoming common objections, educators can foster a positive perception of AI and ensure its successful integration into the classroom. This bonus section has provided strategies and examples to help build trust and maximize the benefits of AI in education. As we move forward, maintaining trust will be essential to tap into the full potential of AI to enhance teaching and learning experiences.

References

- AI's Role in Legal Analysis and Research: A Game Changer. https://www.geekpedia.com/ai-legal-analysis-research-impact/
- Classcraft. (n.d.). Retrieved from https://www.classcraft.com/
- Dialogflow. (n.d.). Retrieved from https://dialogflow.cloud.google.com/
- Facilitating Peer Feedback | Centre for Innovation and Excellence in Learning | Vancouver Island University | Canada. https://ciel.viu.ca/pages/facilitating-peer-feedback
- Family Educational Rights and Privacy Act (FERPA). (n.d.). Retrieved from https://www2.ed.gov/policy/gen/guid/fpco/ferpa/index.html
- General Data Protection Regulation (GDPR). (n.d.). Retrieved from https://gdpr.eu/
- Google Teachable Machine. (n.d.). Retrieved from https://teachablemachine.withgoogle.com/
- Gregorio Ferreira - AI2FUTURE. http://ai2future.com/speakers/gregorio-ferreira/
- IBM Watson Education. (n.d.). Retrieved from https://www.ibm.com/watson/education
- Lavergne, N. (2017). Improve School District-Community Relationships with Social Media. Delta Kappa Gamma Bulletin, 84(2), 13-15,62.
- Masterminds of Mishap: Unveiling the Art and Science of Effective Risk Communication – IzSeo. https://izseo.net/masterminds-of-mishap-unveiling-the-art-and-science-of-effective-risk-communication/
- Microsoft Bot Framework. (n.d.). Retrieved from https://dev.botframework.com/
- Remind. (n.d.). Retrieved from https://www.remind.com/
- Revolutionize Counterterrorism: Unleashing Advanced Technologies to Triumph - Business, Finances, Tech, Markets. https://busilon.com/r

evolutionize-counterterrorism-unleashing-advanced-technologies-to-triumph/
- The Benefits of Bilingual Education: Learning a second language.. https://globalgrowthforum.com/the-benefits-of-bilingual-education-how-learning-a-second-language-can-enhance-cognitive-development/

24

Bonus 3: Spotlight on AI in Education and Human Security

U nderstanding Human Security in the Context of AI in Education

In the context of AI in education, human security extends beyond traditional notions of national security. It emphasizes protecting individuals' health, economic stability, personal safety, and access to education. When integrating AI into the educational sector, it is essential to consider its implications for human security to ensure that deploying these technologies enhances rather than undermines individual and collective safety. The potential benefits of AI in education, such as personalized learning and improved educational outcomes, can inspire optimism and hope in our audience.

Privacy and Data Security

The Importance of Data Protection

AI systems in education rely heavily on data, including personal and sensitive information about students, teachers, and administrative processes. Securing the privacy of this data is paramount. Breaches can result in severe consequences, including fraud, identity theft, cyberbullying, and unauthorized surveillance.

Strategies for Safeguarding Data

1. **Robust Encryption**: Implementing advanced encryption techniques to protect data in transit and at rest.

2. **Access Control**: Enhance security by providing secure access to data to authorized staff only and enable multi-factor authentication methods.

3. **Regular Audits**: Conducting regular and frequent security audits and vulnerability checks to isolate and address potential weaknesses.

Algorithmic Bias and Fairness

Recognizing Bias in AI Systems

AI algorithms can inadvertently perpetuate and amplify existing biases if not carefully designed and monitored. For example, if the data used to train an AI system is biased, the system may make decisions that favor certain groups or individuals over others, leading to discriminatory student admissions, grading, and resource allocation practices.

Ensuring Fairness and Equity

1. **Diverse Data Sets**: Using diverse and representative data sets to train AI models, ensuring they reflect the diverse student population.

2. **Bias Audits**: Regularly auditing AI systems for bias and implementing corrective measures as needed.

3. **Inclusive Development**: Addressing algorithmic bias is not solely the responsibility of developers. Involving stakeholders from various backgrounds in developing and deploying AI systems is crucial. This inclusive approach ensures that diverse perspectives are considered, which can help identify and address potential biases. Such a proactive approach can reassure the audience about the commitment to fairness and equity in AI-driven education.

Psychological and Emotional Well-being

Addressing Mental Health Concerns

AI in education can impact students' mental health in various ways. For instance, constant monitoring and data collection might create a sense of surveillance, leading to anxiety and stress. This emphasizes the importance of mental and emotional well-being in the context of AI in education, evoking empathy and concern in the audience, and highlighting the need for a balanced approach that prioritizes academic and emotional development.

Promoting Positive Mental Health

1. **Transparency and Consent**: Communicating how AI systems collect and use data is a legal requirement and critical to building trust. Obtaining informed consent from students and parents further enhances transparency, making them more confident in the use of AI in education.

2. **Support Systems**: Integrating AI tools with mental health resources to provide timely support and interventions.

3. **Digital Well-being Education**: Educating students about digital well-being and the implications of AI use, empowering them to manage their digital lives responsibly.

Economic Security and Job Displacement

The Impact on Employment

Automating administrative tasks and using AI-driven teaching assistants could lead to job displacement for some educational professionals. Addressing this issue is crucial to maintaining economic security within the education sector.

Mitigating Economic Risks

1. **Reskilling and Upskilling**: Offering training programs to help educators and administrative staff acquire new skills relevant to the evolving educational landscape.

2. **AI Augmentation**: This refers to using AI to enhance human roles rather than replace them. For example, AI can be deployed to automate repetitive tasks, freeing up educators' time for more personalized interactions with students. Thus, it enhances efficiency while maintaining the need for human oversight and interaction.

3. **Economic Policies**: Advocating for policies that support workers affected by AI integration, such as transition assistance and unemployment benefits.

Physical Security and Cybersecurity

Addressing Cyber Threats

As digital tools and platforms are increasingly deployed in educational settings, they become potential targets for cyberattacks, disrupting learning and compromising sensitive information. Physical security concerns in this context can include unauthorized access to AI systems, tampering with AI algorithms, or even physical damage to AI infrastructure.

Enhancing Cybersecurity Measures

1. **Comprehensive Cybersecurity Plans**: Developing and executing robust cybersecurity plans designed to address the specific needs of educational institutions.

2. **Education and Training**: Providing cybersecurity training for students, educators, and staff is crucial. This equips them with the necessary skills to identify and respond to cyber threats and fosters a culture of heightened security awareness, where everyone understands the role of data protection and their part in ensuring it.

3. **Incident Response**: Establishing clear protocols for responding to cyber incidents, minimizing damage, and ensuring rapid recovery.

Conclusion: Prioritizing Human Security in AI-Driven Education

Ensuring human security in the context of AI in education requires a holistic approach that addresses privacy, fairness, mental health, economic stability, and physical security. By implementing comprehensive strategies and fostering a culture of responsibility and transparency, we can harness AI's potential while protecting the well-being of all individuals involved in the educational ecosystem. This proactive stance will help build trust in AI technologies and pave the way for a secure, equitable, and thriving educational future.

References

- Binns, R. (2018). Fairness in machine learning: Lessons from political philosophy. Proceedings of the 2018 Conference on Fairness, Accountability, and Transparency, 149-159. https://doi.org/10.1145/3287560. 3287580

- Buchanan, B. G. (2019). Artificial intelligence in education: Opportunities and challenges. Journal of Learning Analytics, 6(1), 1-17. https://doi.org/10.18608/jla.2019.62.1
- Ensuring Medical Data Security: Safeguarding Patient Information in Wireless Networks Ensuring Privacy in Wireless Sensor Networks: Comprehensive Guide to Data Transmission Techniques. https://wwic 2005.org/ensuring-medical-data-security-safeguarding-patient-information-in-wireless-networks
- Floridi, L., & Taddeo, M. (2016). What is data ethics? Philosophical Transactions of the Royal Society A: Mathematical, Physical and Engineering Sciences, 374(2083), 20160360. https://doi.org/10.10 98/rsta.2016.0360
- Greene, D., Hoffmann, A. L., & Stark, L. (2019). Better, nicer, clearer, fairer: A critical assessment of the movement for ethical artificial intelligence and machine learning. Proceedings of the 52nd Hawaii International Conference on System Sciences, 2122-2131. https://doi. org/10.24251/HICSS.2019.259
- Heaven, W. D. (2020). Our algorithmic overlords: How AI shapes society and what we can do about it. Nature Machine Intelligence, 2(10), 537-538. https://doi.org/10.1038/s42256-020-00246-9
- Horvitz, E. (2017). AI, people, and society. Science, 357(6346), 7. https://doi.org/10.1126/science.aao2466
- Lucifer. (2024, March 20). *5 key elements of a successful data protection strategy*. Eurekafund. https://eurekafund.org/2023/37971/5-key-elements-of-a-successful-data-protection-strategy/
- O'Neil, C. (2016). Weapons of math destruction: How big data increases inequality and threatens democracy. Crown Publishing Group.
- Scherer, M. U. (2016). Regulating artificial intelligence systems: Risks, challenges, competencies, and strategies. Harvard Journal of Law & Technology, 29(2), 353-400. https://doi.org/10.2139/ssrn.2609777

- Selbst, A. D., & Barocas, S. (2018). The intuitive appeal of explainable machines. Fordham Law Review, 87(3), 1085-1139. https://ir.lawnet.fordham.edu/flr/vol87/iss3/6
- Singh, P. K. (2023). Digital Transformation in Supply Chain Management: Artificial Intelligence (AI) and Machine Learning (ML) as Catalysts for Value Creation. https://doi.org/10.59160/ijscm.v12i6.6216
- Tegmark, M. (2017). Life 3.0: Being human in the age of artificial intelligence. Alfred A. Knopf.
- Van Wynsberghe, A., & Robbins, S. (2019). Critiquing the reasons for making artificial moral agents. Science and Engineering Ethics, 25(3), 719-735. https://doi.org/10.1007/s11948-018-0030-8
- Veale, M., Van Kleek, M., & Binns, R. (2018). Fairness and accountability design needs for algorithmic support in high-stakes public sector decision-making. Proceedings of the 2018 CHI Conference on Human Factors in Computing Systems, 440. https://doi.org/10.1145/3173574.3174014
- Whittlestone, J., Nyrup, R., Alexandrova, A., & Dihal, K. (2019). The role and limits of principles in AI ethics: Towards a focus on tensions. Proceedings of the 2019 AAAI/ACM Conference on AI, Ethics, and Society, 195-200. https://doi.org/10.1145/3306618.3314289
- Zuboff, S. (2019). The age of surveillance capitalism: The fight for a human future at the new frontier of power. Public Affairs.

www.ingramcontent.com/pod-product-compliance
Lightning Source LLC
Chambersburg PA
CBHW061140120626
46546CB00005B/1875